D1620071

Edited by
Leszek Moscicki

Extrusion-Cooking Techniques

Further Reading

Brennan, J. G., Grandison, A. S. (eds.)

Food Processing Handbook

2 volumes

2011

Hardcover

ISBN: 978-3-527-32468-2

Peinemann, K.-V., Pereira Nunes, S., Giorno, L. (eds.)

Membrane Technology

Volume 3: Membranes for Food Applications

2010

Hardcover

ISBN: 978-3-527-31482-9

Rijk, R., Veraart, R. (eds.)

Global Legislation for Food Packaging Materials

2010

Hardcover

ISBN: 978-3-527-31912-1

Campbell-Platt, G. (ed.)

Food Science and Technology

2009

Hardcover

ISBN: 978-0-632-06421-2

International Food Information Service

IFIS Dictionary of Food Science and Technology

2009

Hardcover

ISBN: 978-1-4051-8740-4

Janssen, L., Moscicki, L. (eds.)

Thermoplastic Starch

A Green Material for Various Industries

2009

Hardcover

ISBN: 978-3-527-32528-3

Piringer, O. G., Baner, A. L. (eds.)

Plastic Packaging

Interactions with Food and Pharmaceuticals

2008

Hardcover

ISBN: 978-3-527-31455-3

Edited by
Leszek Moscicki

Extrusion-Cooking Techniques

Applications, Theory and Sustainability

WILEY-VCH Verlag GmbH & Co. KGaA

The Editor

Prof. Dr. Leszek Moscicki
Lublin University of Life Sci.
Dept. of Food Proc. Engineer.
Doswiadczalna Str. 44
20-280 Lublin
Poland

Library of Congress Card No.: applied for

British Library Cataloguing-in-Publication Data
A catalogue record for this book is available from the British Library.

Bibliographic information published by the Deutsche Nationalbibliothek
The Deutsche Nationalbibliothek lists this publication in the Deutsche Nationalbibliografie; detailed bibliographic data are available on the Internet at http://dnb.d-nb.de.

Cover Design Formgeber, Eppelheim
Typesetting Thomson Digital, Noida, India
Printing and Binding Fabulous Printers Pte Ltd, Singapore

Printed in Singapore
Printed on acid-free paper

ISBN: 978-3-527-32888-8

Contents

Extrusion-Cooking Techniques: Applications, Theory and Sustainability. Edited by Leszek Moscicki

ISBN: 978-3-527-32888-8

Preface

Extrusion-cooking is gaining increasing popularity in the global agro-food processing industry, particularly in the food and feed sectors. In this handbook, we want to share the secrets of this relatively new technology, paying particular attention to the utilitarian aspects, namely discussing the processes associated with the production of various extrudates and describing the machinery and equipment necessary for their manufacture. There are also many comments and recommendations of a purely operational nature. A perusal of the book will give, in my view, not only information concerning the complexities of the production but also a quantum of engineering knowledge helpful in decision making for those who are considering whether or not to implement this technology in their processing plants.

Being a promoter of substantial research programs in this field for many years I am a great enthusiast for extrusion-cooking, seeing much room for its application. As an editor my ambition was to collect the contents of this book that represent the latest achievements in the field, offering readers the best practical knowledge available at the time of publication. The text of the book has been developed together with invited scientists and industrial experts, whose substantial contribution has assured a high level of transferred knowledge.

This handbook is intended for researchers, students, engineers and technical staff studying or working in food process engineering, food technology, chemical engineering and/or related disciplines. I do hope that the text will also be of interest for food and feed producers as well as representatives of the various industries benefitting from the processing components and semi-finished goods in the agro-food industry.

L. Mościcki - Editor

Extrusion-Cooking Techniques: Applications, Theory and Sustainability. Edited by Leszek Moscicki

ISBN: 978-3-527-32888-8

List of Contributors

Leon P.B.M. Janssen
University of Groningen
Department of Chemical Engineering
Nijenborgh 4
9474 AG Groningen
The Netherlands

Marcin Mitrus
Lublin University of Life Sciences
Department of Food Process
Engineering
Doswiadczalna Str. 44
20-280 Lublin
Poland

Leszek Mościcki
Lublin University of Life Sciences
Department of Food Process
Engineering
Doswiadczalna Str. 44
20-280 Lublin
Poland

Andreas Moster
Bühler GmbH
Ernst-Amme-Str. 19
38114 Braunschweig
Germany

Dick J. van Zuilichem
Everlaan 1
6705 DH Wageningen
The Netherlands

Agnieszka Wójtowicz
Lublin University of Life Sciences
Department of Food Process
Engineering
Doswiadczalna Str. 44
20-280 Lublin
Poland

Extrusion-Cooking Techniques: Applications, Theory and Sustainability. Edited by Leszek Moscicki

ISBN: 978-3-527-32888-8

1
Extrusion-Cooking and Related Technique

Leszek Mościcki and Dick J. van Zuilichem

1.1
Extrusion-Cooking Technology

Extrusion technology, well-known in the plastics industry, has now become a widely used technology in the agri-food processing industry, where it is referred to as *extrusion-cooking*. It has been employed for the production of so-called engineered food and special feed.

Generally speaking, extrusion-cooking of vegetable raw materials deals with extrusion of ground material at baro-thermal conditions. With the help of shear energy, exerted by the rotating screw, and additional heating of the barrel, the food material is heated to its melting point or plasticating point [1, 2]. In this changed rheological status the food is conveyed under high pressure through a die or a series of dies and the product expands to its final shape. This results in very different physical and chemical properties of the extrudates compared to those of the raw materials used.

Food extruders (extrusion-cookers) belong to the family of HTST (high temperature short time)-equipment, capable of performing cooking tasks under high pressure. This is advantageous for vulnerable food and feed as exposure to high temperatures for only a short time will restrict unwanted denaturation effects on, for example, proteins, amino acids, vitamins, starches and enzymes. Physical technological aspects like heat transfer, mass transfer, momentum transfer, residence time and residence time distribution have a strong impact on the food and feed properties during extrusion-cooking and can drastically influence the final product quality. An extrusion-cooker is a process reactor [2], in which the designer has created the prerequisites with the presence of a certain screw lay-out, the use of mixing elements, the clearances in the gaps, the installed motor power and barrel heating and cooking capacity, to control a food and feed reaction. Proper use of these factors allow to stimulate transformation of processed materials due to heating, for example, the denaturation of proteins in the presence of water and the rupture of starches, both affected by the combined effects of heat and shear. These reactions can also be provoked by the presence of a distinct biochemical or chemical component like an enzyme or a pH controlling agent. When we consider the cooking extruder to be more

Extrusion-Cooking Techniques: Applications, Theory and Sustainability. Edited by Leszek Moscicki

ISBN: 978-3-527-32888-8

Figure 1.1 Assortment of popular extrudates.

than just a simple plasticating unit, a thorough investigation of the different physical technological aspects is more than desirable.

Currently, extrusion-cooking as a method is used for the manufacture of many foodstuffs, ranging from the simplest expanded snacks to highly-processed meat analogues (see Figure 1.1). The most popular extrusion-cooked products include:

- direct extruded snacks, RTE (ready-to-eat) cereal flakes and a variety of breakfast foods produced from cereal material and differing in shape, color and taste and easy to handle in terms of production;
- snack pellets – half products destined for fried or hot air expanded snacks, pre-cooked pasta;
- baby food, pre-cooked flours, instant concentrates, functional components;
- pet food, aquafeed, feed concentrates and calf-milk replacers;
- texturized vegetable protein (mainly from soybeans, though not always) used in the production of meat analogues;
- crispbread, bread crumbs, emulsions and pastes;
- baro-thermally processed products for the pharmaceutical, chemical, paper and brewing industry;
- confectionery: different kinds of sweets, chewing gum.

The growing popularity of extrusion-cooking in the global agri-food industry, caused mainly by its practical character, led many indigenous manufacturers to implement it on an industrial scale, based on the local raw materials and supported by detailed economic studies based on the domestic conditions [18]. Extrusion-cooking offers a chance to use raw materials which have not previously displayed great economic importance (e.g., faba bean) or have even been regarded as waste. The domestic market has been enriched with a category of high-quality products belonging to the convenience and/or functional food sector. Of practical importance is the fact that the process in question can be implemented with relatively low effort, does not require excessive capital investment, and most equipment is user-friendly and offers multiple applications.

For easier understanding of the extrusion-cooking technology, as an example, we would like to present the most simple production – direct extrusion of cereal snacks with different shapes and flavors. We will give a general overview of the technological process using a standard set of processing equipment commonly used in such a case.

1.1.1
Preparation of Raw Material

The manner of preparation of the raw material to be processed into food preparations depends upon the ingredients used. In the case of direct extruded snacks this is mainly cereal-based material. Depending on its quality, it must be properly ground and weighed according to the recipe and mixed thoroughly before being fed to the extruder. When conditioning is required, before mixing, water in some quantity is necessarily added for the preparation of the material.

Figure 1.2 presents a diagram of a standard installation for the production of direct extrusion and multi-flavor snacks. In the case of simple maize snacks, that is, not enriched and being single-component products, the processing line is significantly

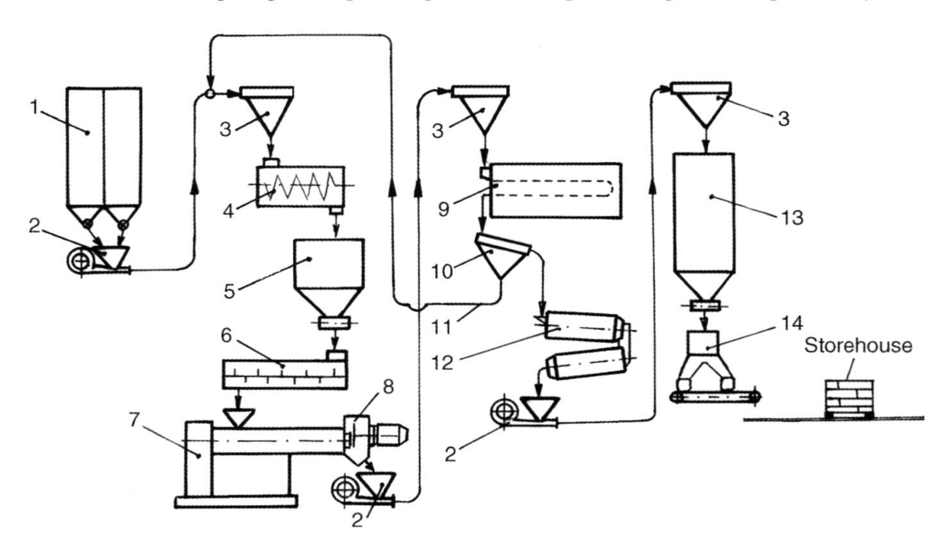

Figure 1.2 A diagram of the set-up for the production of multi-flavor cereal snacks [1]: 1 – a silo with raw materials, 2 – pneumatic conveyer, 3 – collector, 4 – mixer, 5 – weigher, 6 – conditioner, 7 – extruder, 8 – cutter, 9 – dryer, 10 – screen, 11 – recycling of dust, 12 – coating drums, 13 – silos of finished product, 14 – packing machine.

simplified. If is often enough to operate an extruder, for example the one presented

Figure 1.3 Single-screw extrusion-cooker, type TS-45 (designed by L. Moscicki), equipped with an electric heating system and a water-air cooling system [1].

in Figure 1.3, and a packing machine to initiate production (often called "garage box production").

1.1.2 Extrusion-Cooking

The effect of direct extrusion-cooking is that, after leaving the die, the material expands rapidly and the extrudates are structurally similar to a honeycomb, shaped by the bundles of molten protein fibers. In this case a simple, single-screw food extruder can be used to manufacture various types of products, different in shape, color, taste and texture [1, 3–5]. The technology for each of them requires an appropriate distribution of temperature, pressure and moisture content of the material during processing. Because the main task is to obtain good-quality extrudates, flexibility and precise control, especially of the thermal process, is essential in the design and construction of modern cooking extruders. More than often, the process for the manufacture of specific products has to be developed empirically.

Particularly interesting are the issues related to the energy consumption of extrusion-cooking of vegetable raw materials. There is a widespread opinion that this power consumption is too high. It is not clear to us on what these opinions are based, since the results of our own research and of those available from the literature mention something completely opposite. Measurements of energy consumption in single-screw food extruders are in the range 0.1–0.2 kWh kg^{-1} (excluding of course, the costs of material preparation, that is, the grinding and conditioning) [1]. This demonstrates that extrusion-cooking is highly competitive in comparison with the conventional methods of thermal processing of vegetable material. Of course, this does not mean that extrusion-cooking is ideal for all applications. It is an alternative and, in many cases, competitive in relation to other methods of food and feed manufacture.

1.1.3 Forming, Drying and Packing

Depending on the purpose for which they are to be used, extrudates must be suitably formed. The melt mass leaving the extruder takes more or less the shape of the extruder-dies(nozzle); at the same time a lengthwise arrangement – an appropriate setting of the speed of a rotary knife cutter installed outside the die, controls the product length. This allows the production of miscellaneous shapes of extrudate such as balls, rings, stars, letters of the alphabet, and so on.

The next stage of production is the drying of the extrudates to a moisture content of about 6–8% and subsequent cooling. Drying can be performed with simple rotating drums with electric heaters installed or with a gas-operated hot air installation working at temperatures just above 100 °C. In larger installations belt dryers are used, heated by gas fired heat exchangers or steam, where the air is circulating through the unit sections. Cooling takes place at the ambient temperature of 15–20 °C, where the air flows through the perforated belt of the dryer–cooler device.

Very often in drum dryers the flavor and vitamin-coating step is integrated (Figure 1.4). The selection of sprinklers and flavor additives is wide: from smoked meat flavor to peanut butter aroma.

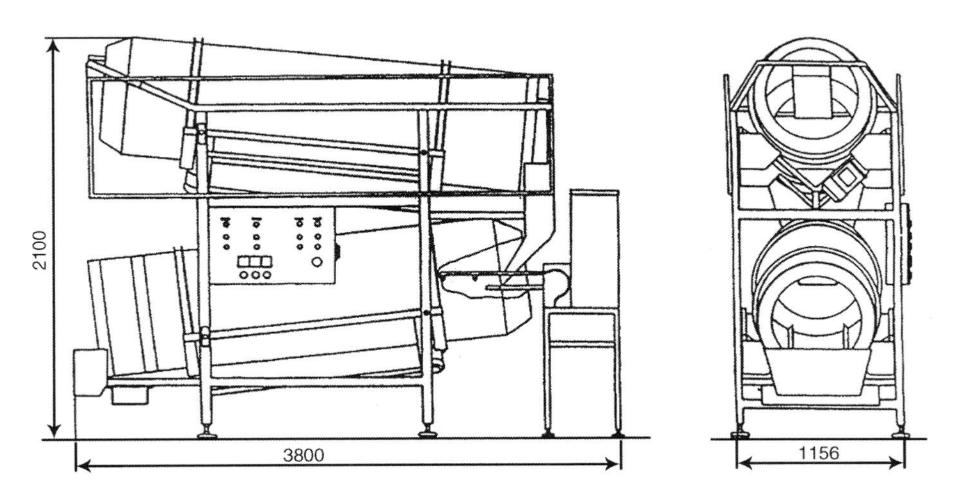

Figure 1.4 Drum dryer and coating drum unit [6].

Direct extrusion snacks in an icing sugar coating are very popular. For small-scale production, it is sufficient to operate a drop coating machine (remember about suitable tempering of the products). Industrial production of this type of cereal product requires the use of an additional high-performance drum dryer or belt dryer, so that the application of the coating can be a continuous process while maintaining a fixed product quality. Some noteworthy examples of the products in question are coated cereal balls, rings or shells, offered by many breakfast cereals producers.

1.2
Quality Parameters

From the foregoing it can be stated that a food extruder may be considered as a reactor in which temperature, mixing mechanism and residence time distribution are mainly responsible for a certain physical state, as is the viscosity. Quality parameters such as the texture are often dependent on the viscosity. The influence of various extruder variables like screw speed, die geometry, screw geometry and barrel temperature on the produced quality has been described by numerous authors for many products [2, 5, 7–9, 18, 20, 21]. However, other extrusion-cooking process variables like initial moisture content, the intentional presence of enzymes, the pH during extrusion, and so on, also play a role. Although a variety of test methods is available a versatile instrument to measure the changes in consistency during pasting and cooking of biopolymers is hardly available. A number of measuring methods are used in the extrusion-cooking branch. A compilation of them is given in Figure 1.5, from which it can be seen that an extrusion-cooked product is described in practice by

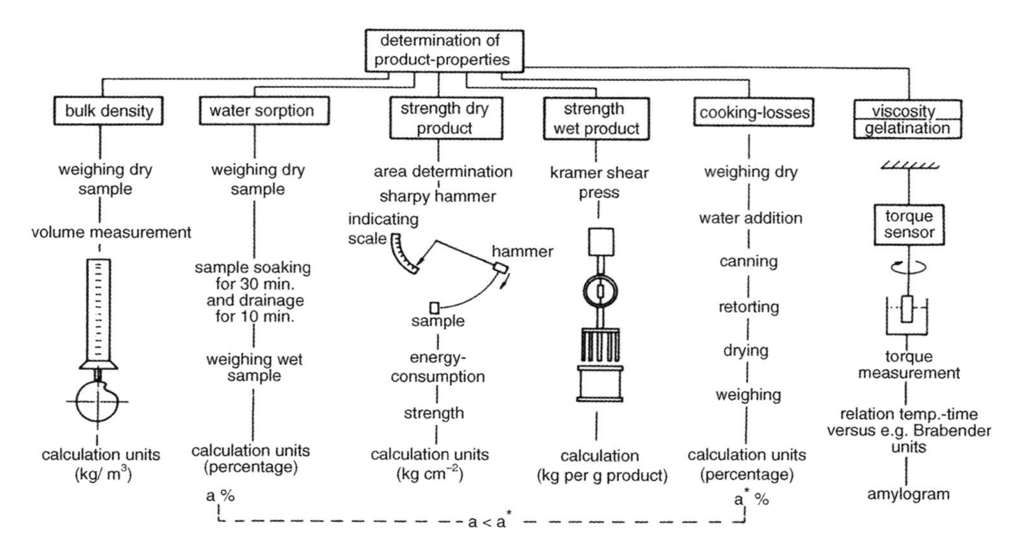

Figure 1.5 Measuring methods for extruder product properties [7].

its bulk density, its water sorption, its wet strength, its dry strength, its cooking loss and its viscosity behavior after extrusion for which the Brabender viscometer producing amylographs may be chosen, which gives information about the response of the extruded material to a controlled temperature–time function (Figure 1.5).

For a successful description of properties of starches and proteins we will need additional chemical data like dextrose-equivalents, reaction rate constants and data describing the sensitivity to enzymatic degradation. The task of the food engineer and technologist will be to forecast the relation between these properties and their dependence ofnthe extruder variables. Therefore, it is necessary to give a (semi)-quantitative analysis of the extrusion- cooking process of biopolymers, which can be done by adopting an engineering point of view, whereby the extruder is considered to be a processing reactor. Although the number of extrusion applications justify an optimistic point of view, more experimental verification is definitely needed, focussed on the above-mentioned residence time distribution, the temperature distribution, the interrelation with mechanical settings like screw compositions, restrictions to flow, and so on.

1.3 Extrusion-Cooking Technique

As was mentioned already, extrusion-cooking is carried out in food extruders – machines in which the main operative body is one screw or a pair of screws fitted in a barrel. During baro-thermal processing (pressure up to 20 MPa, temperature 200 °C), the material is mixed, compressed, melted and plasticized in the end part of the machine (Figure 1.6). The range of physical and chemical changes in the processed material depends principally on the parameters of the extrusion process and the construction of the extruder, that is, its working capability.

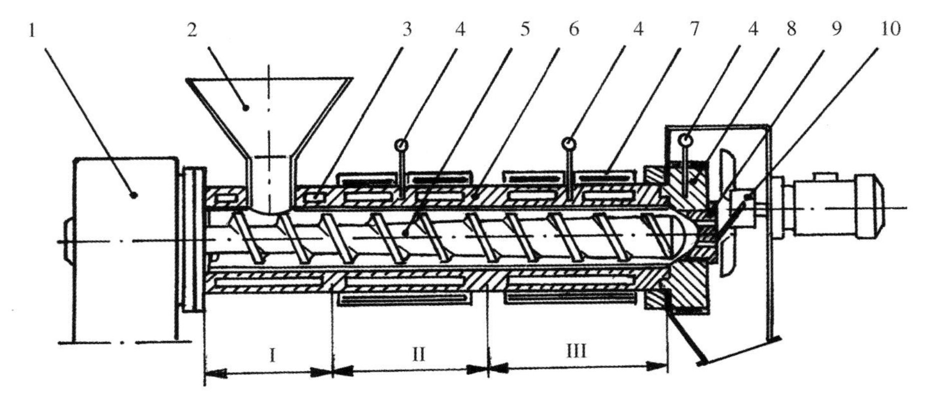

Figure 1.6 A cross-section of a single-screw food extruder: 1 – engine, 2 – feeder, 3 – cooling jacket, 4 – thermocouple, 5 – screw, 6 – barrel, 7 – heating jacket, 8 – head, 9 – dies, 10 – cutter, I – transport section, II – compression section, III –melting and plasticizing section [1].

There are many conventional methods of classification of food extruders but, in our opinion, the most practical is the one taking into account the following three factors.

1) The method of generating mechanical friction energy converted during extrusion into heat (three types of extruders):
   a) autogenic (source of heat is the friction of the particles of the material caused by the screw rotating at high speed);
   b) isothermic (heated);
   c) polytropic (mixed).
2) The amount of mechanical energy generated (two types of extruders):
   a) low-pressure extruders producing relatively limited shear rate;
   b) high-pressure extruders generating large amounts of mechanical energy and shear.
3) The construction of the plasticizing unit (see Figure 1.7), where both the barrel and the screw may be designed as a uniform, integrated body or fixed with separate modules.

1.3.1
Historical Development

At first, in 1935, the application of single-screw extruders for plasticating thermoplastic materials became more common as a competitor to hot rolling and shaping in hydraulic-press equipment. A plasticating single-screw extruder is provided with a typical metering screw, developed for this application (Figure 1.8).

In the mid-1930s we notice the first development of twin-screw extruders, both co-rotating and counter-rotating, for food products. Shortly after, single-screw extruders came into common use in the pasta industry for the production of spaghetti and macaroni-type products. In analogy with the chemical polymer industry, the single-screw equipment was used here primarily as a friction pump, acting more or less as continuously cold forming equipment, using conveying-type screws. It is remarkable that nowadays the common pasta products are still manufactured with the same single-screw extruder equipment with a length over diameter ratio (L/D) of approximately 6–7. However, there has been much development work on screw and die design and much effort has been put into process control, such as sophisticated temperature control for screw and barrel sections, die tempering. and the application of vacuum at the feed port. Finally, the equipment has been scaled up from a poor hundred kilos hourly production to several tons [1, 2].

The development of many different technologies seems to have been catalyzed by World War II, as was that of extrusion-cooking technology. In 1946 in the US the development of the single-screw extruder to cook and expand corn- and rice-snacks occurred. In combination with an attractive flavoring this product type is still popular, and the method of producing snacks with single-screw extruder equipment is, in

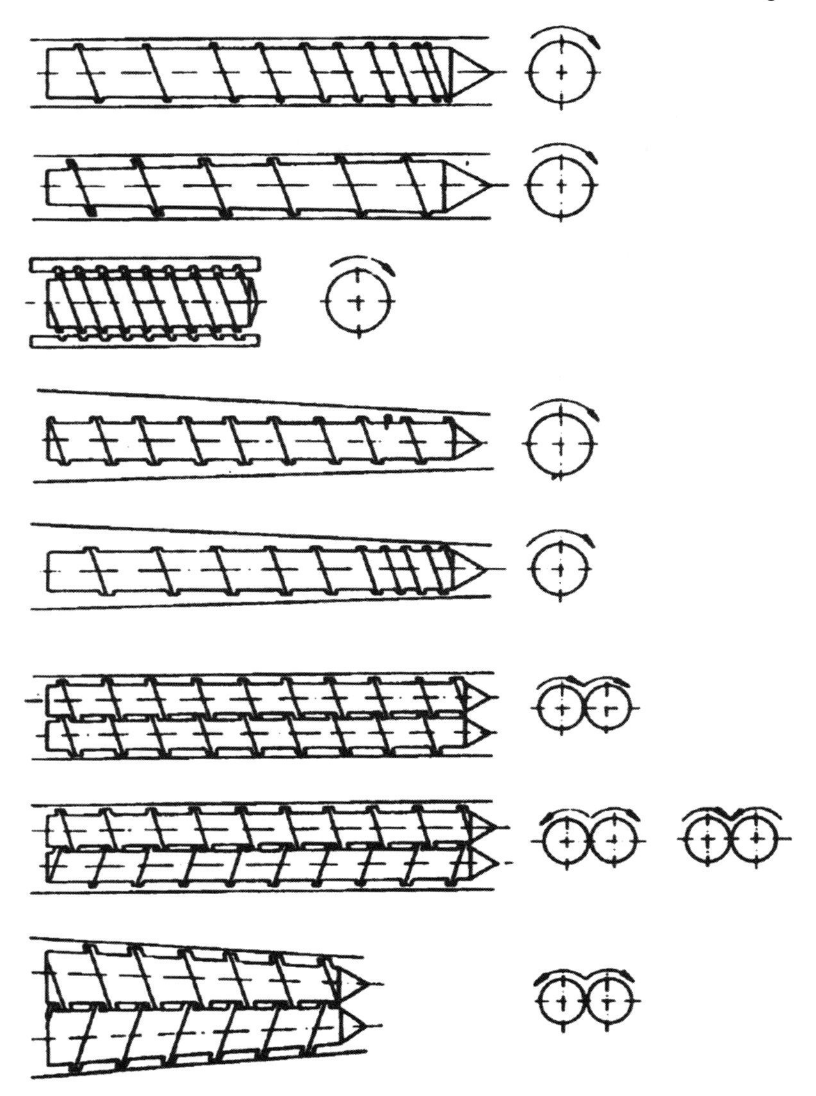

Figure 1.7 Configurations of screws' geometry in the extruder [7].

principle, still the same. A wide variety of extruder designs is offered for this purpose. However, it should be mentioned that the old method of cutting preshaped pieces of dough out of a sheet with roller-cutters is still in use, because the complicated shapes of snacks lead to very expensive dies and die-heads for cooking and forming extruders. Here, the lack of knowledge of the physical behavior of a tempered dough and the unknown relations of the transport phenomena of heat, mass and momentum to the physical and physico-chemical properties of the food in the extruder are clearly noticed. Although modern control techniques are very helpful in controlling

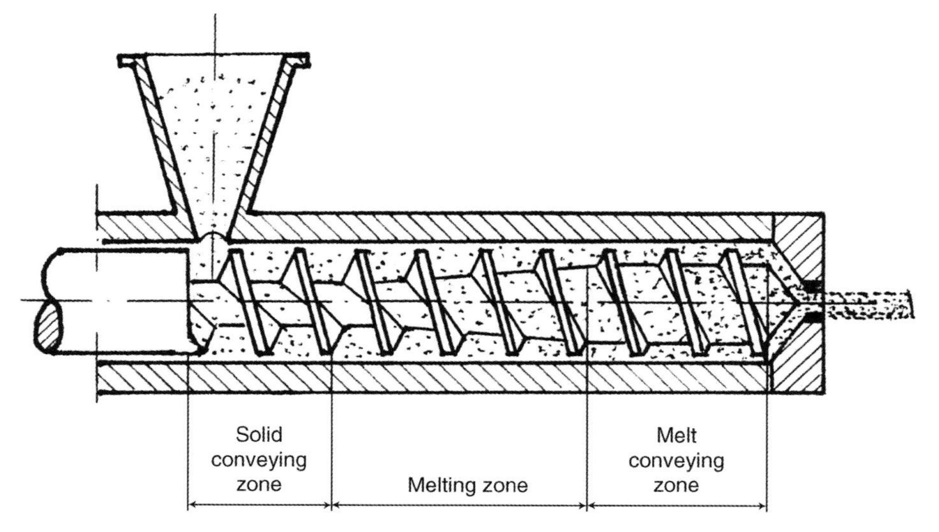

Figure 1.8 Typical plasticating single screw extruder [2].

the mass flow in single-screw extruders, in many cases it is a big advantage to use extruders with better mixing and more steady mass flow than single screw equipment can offer.

In the mid-1970s the use of twin-screw extruders for the combined process of cooking and forming of food products was introduced, partly as an answer to the restrictions of single-screw extruder equipment since twin-screw extruders provide a more or less forced flow, and partly because they tend to give better results on scale up from the laboratory extruder types in use for product development [2].

1.3.2
Processing of Biopolymers

When we focus on food we notice that nearly all chemical changes in food are irreversible. A continued treatment after such an irreversible reaction in an extruder should be a temperature-, time- and shear-controlled process leading to a series of completely different functional properties of the produced food (Table 1.1).

Nowadays, food familiar extruder equipment manufacturers design process lines, where the extruder-cooker is part of a complete line. Here, the extruder is used as a single- or twin-screw reactor, and the preheating/preconditioning step is performed in a specially built preconditioner. The forming task of the extruder has also been separated from the heating and shearing. The final shaping and forming has to be done in a second and well optimized post-die forming extruder, processing the food mostly at a lower water level than in the first reactor extruder. In such a process line cooked and preshaped but unexpanded food pellets can be produced (see Chapters 4 and 5). The result is a typical food process line differing very much from a comparable extruder line in the chemical plasticizing industry.

Table 1.1 Comparison between thermoplastic-polymers and biopolymers [2].

		Plastics	Food
1	Feed to the extruder	Single polymer	Multiple solids, water and oil
2	Composition	Well defined structure and molecular weight	Not well defined. Natural biopolymers, starch, protein, fiber, oil and water
3	Process	Melting and forming. No chemical change. Reversible	Dough or melt-like formation with chemical change. Irreversible continuous treatment leads to wanted specific functional properties
4	Die forming	Shape is subjected to extrudate swell	Subjected to extrudate swell and possibly vapor pressure expansion
5	Biochemicals	Use of fillers, for example, starch	Use of enzymes and biochemicals for food conversion

With the use of the extrusion-cooking equipment in process lines their tasks became more specialized. This encouraged the comparison of food extruder performance with that in the plastics industry, thus promoting the transformation of the extrusion-cooking craft into a science, tailor-made for the "peculiar" properties of biopolymers.

1.3.3 Food Melting

If the aim of the extrusion-cooking process in question is a simple denaturation of the food polymer, without further requirements of food texture, then the experience of the chemical polymer extrusion field, applying special screw melting parts, is advisable, as the melting will be accelerated. In principle, the effect of these melting parts is based on improved mixing. This mixing effect can be based on particle distribution or on shear effects exerted on product particles. For distributive mixing the effects of mixing are believed to be proportional to the total shear γ given by:

$$\gamma = \int_0^t \frac{dv}{dx} dt \tag{1.1}$$

Whereas for dispersive (shear) mixing the effect is proportional to the shear stress τ:

$$\tau = \mu \frac{dv}{dx} \tag{1.2}$$

The group of distributive mixing screws can be divided into the pin mixing section, the Dulmage mixing section, the Saxton mixing section, the pineapple mixing head, slotted screw flights, and the cavity transfer mixing section, respectively [2, 17]. The pin mixing section, or a variant of this design, is used for food in the Buss co-kneader

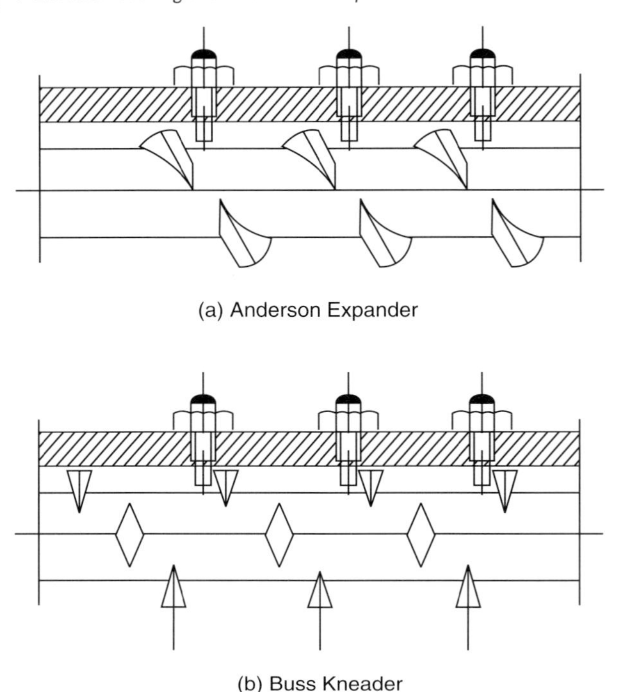

Figure 1.9 Mixing pin elements Anderson and Buss designs [2].

which is a reciprocating screw provided with pins of special design, rotating in a barrel also provided with pins Figure 1.9). When we recognize food expanders to be food extruders then much use is made here of the pin-mixing effect, since the barrels of the expander equipment, built like the original Anderson design, are provided with mixing pins (see Chapter 11). The amount of mixing and shear energy is simply controlled by varying the number of pins. Some single-screw extruder manufacturers have designs available, like, for example, special parts of the Wenger single-screw equipment, where some influence of the mentioned designs is recognizable. The pineapple mixing section or the most simple mixing torpedo has a future in food extrusion cooking due to its simplicity and effectiveness.

1.3.4 Rheological Considerations

It has already been mentioned that non-Newtonian flow behavior is usually to be expected in food extruders. A major complication is that chemical reactions also occur during the extrusion-cooking process (e.g., gelatinization of starch or starch-derived materials, denaturation of proteins, Maillard reactions), which strongly influence the viscosity function. The rheological behavior of the product, which is relevant to the modeling of the extrusion process, has to be defined directly after the extruder screw tip, before expansion has occurred, in order to prevent the influence of water losses, cavioles in the material and temperature effects due to the flashing process that occur as soon as the material is exposed to the environment. A convenient way would be to measure the pressure loss over capillaries of variable diameter and length. However, this method has a certain lack of accuracy, since pressure losses due to entry effects are superimposed on the pressure gradient induced by the viscosity of the material [10]. Moreover, since the macromolecules in the biopolymers introduce a viscoelastic effect, the capillary entry and exit effects cannot be established easily from theoretical considerations.

It is well known that within normal operating ranges starches and protein-rich materials are shear thinning. This justifies the use of a power law equation for the shear dependence of the viscosity [19]:

$$\eta_a = k'|\dot{\gamma}|^{n-1} \tag{1.3}$$

where η_a is the apparent viscosity, $\dot{\gamma}$ is the shear rate and n is the power law index.

Metzner [11] has argued that for changing temperature effects this equation can be corrected by multiplying the power law effect and the temperature dependence, thus giving:

$$\eta_a = k'|\dot{\gamma}|^{n-1} \exp(-\beta \Delta T) \tag{1.4}$$

The viscosity will also be influenced by the processing history of the material as it passes through the extruder [12]. In order to correct for this changing thermal history one should realize that interactions between the molecules generally occur through the breaking and formation of hydrogen and other physico-chemical bonds. This cross-linking effect is dependent on two mechanisms: a temperature effect determines the frequency of breaking and the formation of bonds and a shear effect determines whether the end of a bond that breaks meets a "new" end or will be re-attached to its old counterpart. If we assume that this last mentioned effect will not be a limiting factor as soon as the actual shear rate is higher than a critical value and that the shear stress levels within the extruder are high enough, then the process may be described by an Arrhenius model, giving, for the reaction constant [16]:

$$K(t) = k_\infty \exp - \frac{\Delta E}{RT(t)} \tag{1.5}$$

where ΔE is the activation energy, R the gas constant and T the absolute temperature. Under the assumption that the crosslinking process may be described as a first order reaction, it is easy to show that from the general reaction equation:

$$\frac{dC}{dt} = K(1-C) \tag{1.6}$$

can be derived:

$$1-C = \exp\left(- \int_0^{\tau} \exp - \frac{\Delta E}{RT(t)} Dt \right) \tag{1.7}$$

in which C denotes the ratio between actual crosslinks and the maximum number of crosslinks that could be attained, and where Dt is a convective derivative accounting for the fact that the coordinate system is attached to a material element as it moves through the extruder. Therefore, the temperature, which is of course stationary at a certain fixed position in the extruder, will be a function of time in the Lagrangian frame of reference chosen. This temperature history is determined by the actual position of the element in the extruder, as has been proved for synthetic polymers by Janssen *et al.* [13]. It is expected that this effect will cancel out within the measuring accuracy and that an overall effect based on the mean residence time τ may be chosen. It may now be stated that the apparent viscosity of the material as it leaves the extruder may be summarized by the following equation:

$$\eta_a = k''|\dot{\gamma}|^{n-1}\exp(-\beta T)\exp\left(\int_0^{\tau}\exp-\frac{\Delta E}{RT(t)}\,Dt\right) \tag{1.8}$$

Under the restrictive assumption that k, n, β and ΔE are temperature independent, it is obvious that at least four different measurements have to be carried out in order to characterize the material properly [2, 12]

1.4 Modern Food Extruders

1.4.1 Single-Screw Extrusion-Cookers

As mentioned before the design of single-screw extrusion-cookers is relatively simple The role of the screw is to convey, compress, melt and plasticize the material and to force it under pressure through small die holes at the end of the barrel. The necessary condition for moving the material is a proper flow rate and no sticking to the surface of the screw. In food extruders sticking effects are prevented by the force of friction of the material against the barrel wall, which is facilitated by suitable grooving of the inside of the barrel (longitudinal or spiral grooves). Their role is to increase the grip-resistance and to direct the flow of the forced material. The principle is the following: the more friction, the less spinning of the material and easier transport forward.

Single-screw food extruders process relatively easy materials characterized by a high friction coefficient, such as maize or rice grits. Such grits can be extruded even under a pressure of around 15–20 MPa, and are basic materials for the production of direct extrusion snacks or breakfast cereals (balls, rings, etc.). For these materials, even the use of the simplest autogenic extruders is sufficient; with a low ratio of screw length L and diameter D (L/D = 4–6; see Figure 1.10). Unfortunately, the main disadvantage of single-screw extruders is poor mixing of the material. This should be done before feeding. Also, single-screw extruders show a limited efficiency, especially when multi-component mixtures of raw materials are used.

Figure 1.10 Autogenic extruder, type 90 E (permission of Lalesse-Extrusion BV).

Smooth positive movement of the material in a single-screw food extruder depends on the actual drag flow, caused by the screw geometry and its rotation, minus the so-called back flow [17, 19]. In order to maintain the correct working point of the extrusion-cooking process it is necessary to follow strictly the technological regime, that is, to maintain the relevant working parameters of the machine, determined experimentally, or given by expert technicians. The proper preparation of raw materials, especially their grinding and moisturizing, is also important. Often preconditioners are installed additionally (Figure 1.11).

There are a number of points that are useful in daily extruder operations:

- Moisture content and particle size distribution of raw material mixtures must be homogeneous – this will prevent irregular work of the extruder (shooting or blocking) and will ensure the desired quality of the extrudates.

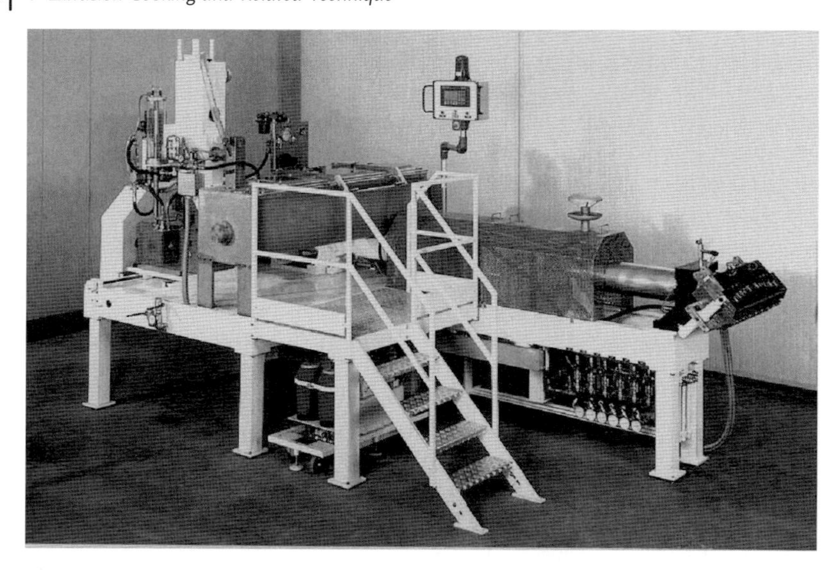

Figure 1.11 Modern single screw extrusion-cooker G-type equipped with additional operating devices (permission of Pavan Group).

- Reduced moisture content of raw materials influences the pressure of extrusion but does not have an essential impact on the extruder's performance (i.e., increase in the viscosity of the material).
- Intensive cooling of the barrel (e.g., with cold water) contributes to a lower temperature and increases the friction inside the material. It must be correlated with the quality requirements of the extrudate. Temperature drop in the material raises its viscosity and contributes positively to the extruder's performance.
- The blocking of a few die-holes results in a sudden increase in pressure and leads to a powerful back flow, or even lockout of the machine. In such a case it is worth trying to clean them immediately with a thin tool, and, if this does not help, then stop the machine and dismantle the die. To postpone the disassembly of the die with a plasticized material inside may lead to permanent damage to the equipment during the next start-up.
- The smallest holes in the die, cause a higher resistance during the extrusion of the material; since small openings increase pressure and reduce the extruder's output as the back flow is higher.
- By applying a plasticizing screw of greater L/D ratio, it is possible to generate more extrusion pressure due to a longer fully filled screw, which leads to a better plasticizing of the material and a reduced back flow.
- The loss of a small clearance between the barrel surface and the screw flights (in practice just 0.1–0.2 mm is enough) hampers the movement of the material, reduces friction and pressure and causes poor operation of the machine and stops the flow, which means poor product or no product at all.

1.4.2 Twin-Screw Extrusion-Cookers

Twin-screw food extruders are much more complex and more universal in terms of design. They have gained extensive popularity with producers of extrusion-cooked food and feed because of their high versatility (the capability of processing a wider range of materials, including viscous and hard-to-break materials), lower energy consumption and the ability to broaden the production assortment significantly. Their only disadvantage is the more complicated design and the cost of acquisition.

Nowadays co-rotating food extruders are used to a greater extent (Figure 1.12) due to their high productivity, good mixing and high screw speed (up to 700 rpm). They are characterized by good efficiency of the material transportation, mixing, plasticizing and extrusion. The self-wiping and intermeshing flights of the screws effectively force the material to move forward, in effect, no material is locked in the space between the surface of the barrel and the screw. For this reason, twin-screw extrusion-cookers are often referred to as self-cleaning machines.

The flow of mixed material in co-rotating twin-screw extruders is balanced without any discontinuity and no C-shaped chambers or prints of characteristic angular waves on the surface of products [15]. This is a decisive factor in the use of these extruders for the production of crispbread or sponge fingers, that is, products with a higher quality of external surface.

The description of physical processes associated with the mechanisms of material transfer in twin-screw food extruders and the accompanying heat exchange is more complicated than in single-screw extruders. More on this subject is given in Chapter 2. Readers who wish to learn more about the engineering aspects of extrusion will certainly be able to master the art of day-to-day handling not only of extruders but also complete processing lines. We use the word "art" deliberately, for the extrusion-cooking technique is fairly complicated and its proper use requires considerable expertise of the operators. The experience and observations gained in a regular production-surrounding are of even greater value than the results of scientific research. Even today, in many cases, we have to admit that progress in the method of producing many assortments of extrudates, and also the practical

Figure 1.12 Modern twin-screw extruder, type BCTA (permission of Bühler AG).

effects of the technologists' work, went on ahead of a detailed scientific description of the physical and chemical phenomena of the process. This shows the high potential of the extrusion-cooking technique, which has become a tool in the hands of the users themselves in their persistent wish to continuously renew their assortment of products.

Counter-rotating twin-screw food extruders are special-purpose machinery (Figure 1.13). Their screws rotate much moreslowly (up to 150 rpm) but can mix the material effectively, and their work resembles a positive-displacement pump generating high pressure in the barrel closed C-shaped chamber on the screws, which is needed for high viscosity material. The back flow of material in these extruders is very small due to the tiny clearances between the screws and the barrel. They are predominantly used for the production of confectionery, chewing gum, and for the processing of fiber and cellulose-rich materials. Counter-rotating extruders can easily be degassed. To use them for the manufacture of simple forms of extrudates would be uneconomic and energy consuming. This does not mean that

Figure 1.13 Counter-rotating twin-screw extrusion-cooker, type Valeurex, of modular construction; the brainchild of Polish, Dutch and Swedish designers cooperating through the European Programme Eureka [14].

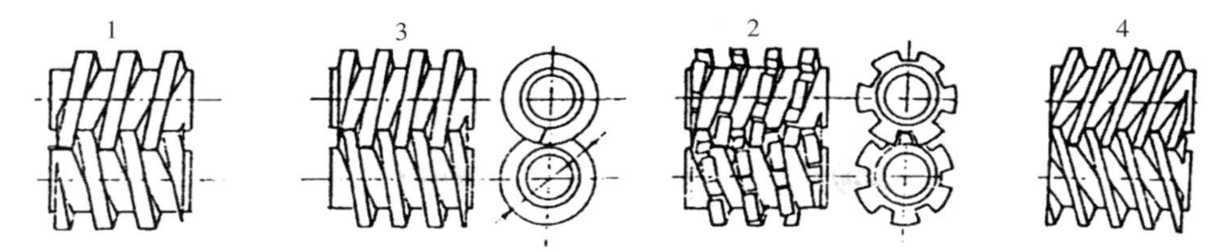

Figure 1.14 A review of screw elements: 1 – transporting elements, 2 – mixing elements, 3 – double flight elements, 4 – compressing elements.

they should not be used for the production of popular multi-component extrudates. These types of machines are successfully used by Polish producers of crispbread, fiber-rich extrudates and even pet food (Figure 6.2).

Modern twin-screw food extruders are designed in such a way that raw materials can be fed to the extruder by more than one feeder, even at different locations through the barrel. Now fluid components can be fed separately, which is an additional advantage. The basic material is precisely fed into the barrel with single and twin-screw feeders. Gravimetric feeders in the form of a vibrated feed tray have practically disappeared, due to the problems with irregular proportioning of the material. Very often in the case of breakfast cereals or snack pellets production, mixtures of raw materials are additionally steam-treated, before extrusion, in suitable pre-conditioners and/or specially designed mixers.

Nowadays twin-screw food extruders have a modular construction where screws are built up out of several different elements, mounted on the screw shaft. These elements are handling transport, mixing, compressing, melting. By proper setting of the elements the operator is able to "control" the behavior of material inside the extruder, and influence the scope of physical and chemical processes in the course of an extrusion-cooking process (Figures 1.14–1.16).

The selection of particular elements and arranging a screw requires relatively broad experience and knowledge of the production targets. Such sophisticated

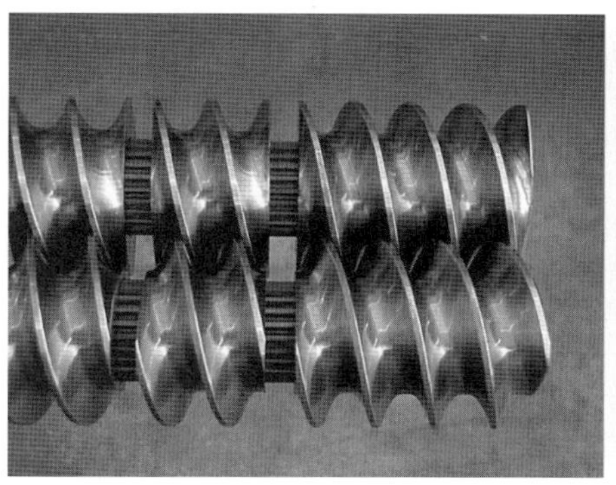

Figure 1.15 A set-up of screw elements and configured screws (with permission of the Pavan Group).

Figure 1.16 A set-up of screws in a modern twin-screw extrusion-cooker equipped with an axial opening barrel (permission of the Pavan Group).

extruders can be very useful production tools but only in the hands of conscientious and experienced operators. Otherwise, they will only be very expensive "toys" whose purchase is not economically justified. Where production is limited to one or two relatively easy products a more cost-effective solution would be to use simpler single-screw food extruders.

Popular co-rotating twin-screw food extruders used for the production of RTE breakfast cereals or pet food run with a screw speed of around 300 rpm. Some machines can operate at a speed which is two or even three times higher. This gives the highest possible production efficiency but the application of special construction materials increases the cost. Extruders often produce a wide range of products, ranging from simple maize snacks to protein-based food. For snacks high pressure and mechanical energy at low L/D ratio is needed; for protein material long processing times and many intermediate stages will be applied. Modular screws give these possibilities, either by the disassembly of the barrel units (for example, 3 out of 6) combined with the replacement of the screws with shorter ones (Figure 1.12), or by changes in the screw lay-out. An example is the elongation of the screw transport section by mounting more transport elements at the cast of other elements on the splinted shafts, in the case of fully opening extruders (Figure 1.16). Several extruder makers offer the use of a mobile feeder to be installed on the feed ports between the center and the end of the barrel, for example, maize grits fed to a possible unlocked opening with a mobile feeder on the arm – see Figure 1.17. In such cases, distance rings are placed on the screw shafts ahead of the final set of screw elements.

During the operation of extruders it is very important to maintain the engine load at an adequate level (max. 80–85%). This means that rpm, torque and energy

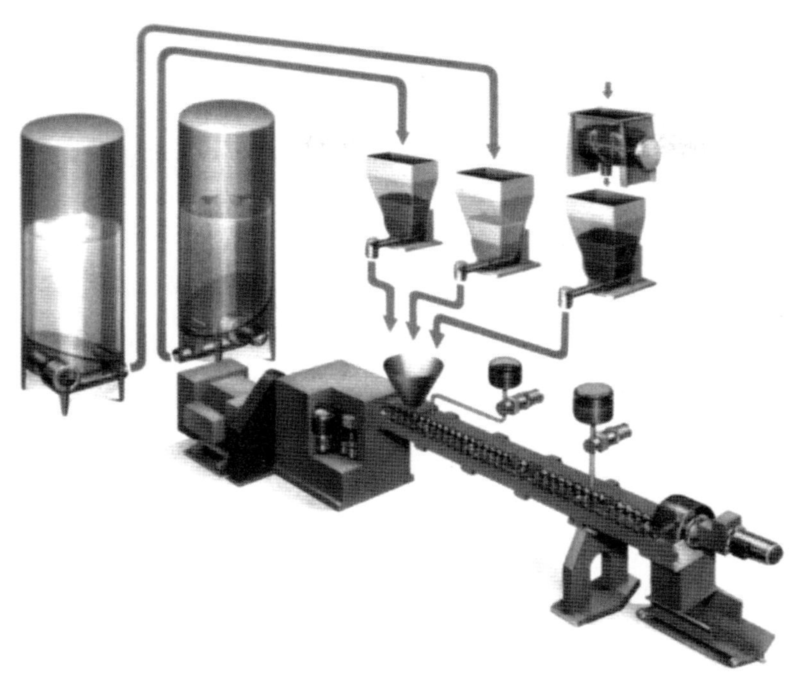

Figure 1.17 Section feed of material and fluid components [1].

consumption, that is SME rate expressed in kWh per 1 kg of the product, must be controlled. These issues are of vital importance since the manufacture of food extrudates necessitates high energy use. An appropriate choice of extruder machinery for the type of production determines the return on investment. As seen from production practice, common opinions about the high energy consumption of extrusion are not true, unless the machinery is not properly utilized.

At this point is also worth noting that there are no established and strict recommendations concerning, for example, the range of temperature in different zones of the plasticizing section of extruder-cookers. We can certainly mention some temperature ranges to be recommended during the processing of certain raw materials but, in the case of a particular machine, this should be verified empirically or the manufacturer's recommendations should be followed. Unfortunately, many manufacturers do not have such knowledge or do not want to share it with extruder users. The distribution of temperature sensors and the accuracy of their readings depend on their mounting depth and distance from the immediate zone of the material flow inside the barrel. This determines approximately the history during processing because we are not able to read the temperature of the material directly inside the barrel. The thermocouple readings represent the estimated temperature of the material; however, extruder manufacturers vary in terms of construction and installation of temperature measurement devices. The same is true about identifying the optimal torque and load of the power transmission system.

In such a case, the operator's experience and the manufacturer's recommendations are of utmost value.

In the daily operation of food extruders, the start-up and close-down of production causes difficulties. This is a particular job for the operator because it is relatively easy to block the machinery; patience and time should be taken as after stopping the plasticizing unit must be cleaned. If the stop procedure is incorrect permanent damage to the equipment can occur.

The start-up procedure is more or less uniform. After heating up and reaching the desired temperature, in individual zones of the plasticizing section, the engine is started by setting its rpm to 1/3 of the nominal level and then highly moisturized material is fed. Subsequently, the machine performance is increased step by step, closely watching the outflow of material and engine load. After a few minutes, the machine reaches its nominal operation parameters.

The lockout must be done in the reverse order. However, remember not to turn off the heaters too early and maintain the system temperature at about 100 °C until the end of the shut-down procedure. After stopping the screws, the die-head is taken down and the machine run again while feeding some coarse material (for example, oats) for the final cleaning of the plasticizing section. The machine will not restart when the working parts and the head are dirty.

The latest state of the art is that the control of the production process is fully automated. In the case of high-volume production rates, this is indispensable, since only a properly programmed control system is fast enough to control the production flow as it responds immediately to failures of the working equipment. These issues will be dealt with in Chapter 13.

1.5 Concluding Remarks

In conclusion it can be said that there is still a strong impact of plastic polymer extrusion on the field of food extrusion technology. First, there is the availability of well developed and refined hardware, including extruder equipment and instrumentation and control systems. They still have to be developed further and/or sifted out to suit specific food applications. Secondly, there is the limited availability of polymer engineering process know-how, since this know-how is company bound. However, it has formed the basis of control techniques in food extrusion-cooking.

The extrusion-cooker is still a relatively new piece of equipment and specialists in that field expect the food extruder to be a process tool capable of helping the industry to develop new series of products. For this purpose one can make use of the unique property of the extrusion-cooker to be a high temperature/pressure short residence time (HTST) piece of equipment, capable for example, of replacing conventional process lines.

The market expects new food products: fancy in shape, taste and raw material composition as well as attractive from an economic point of view. Extrusion-cooking technology can meet these expectations, however one needs specialized knowledge.

Notation

C	Concentration ($kg\,m^3$)
Dt	Convective derivative
k	Consistency factor of power-law model ($N\,s^n\,m^{-2}$)
k_∞	Frequency factor (s^{-1})
$K(t)$	Reaction constant (s^{-1})
n	Flow behavior index of power-law model
p^*	Power number
R	Gas constant ($kJ\,mol^{-1}\,K^{-1}$)
Re	Reynolds number
T	Temperature (K)
v	Velocity ($m\,s^{-1}$)
x	Channel depth (m)
β	Temperature correction constant
ΔE	Activation energy ($kJ\,mol^{-1}$)
γ	Shear
$\dot{\gamma}$	Shear rate (s^{-1})
η	Apparent viscosity ($N\,s\,m^{-2}$)
μ	Dynamic viscosity ($N\,s\,m^{-2}$)
τ	Shear stress ($N\,m^{-2}$)

References

1 Mościcki, L., Mitrus, M., and Wojtowicz, A. (2007) *Technika ekstruzji w przetwórstwie rolno-spożywczym (in Polish)*, PWRiL, Warszawa.

2 Van Zuilichem, D.J. (1992) Extrusion Cooking. Craft or Science? Ph.D. thesis, Wageningen University, Netherlands.

3 Guy, R. (2001) *Extrusion Cooking, Technologies and Applications*, CRC Press Inc., Boca Ration, FL.

4 Harper, J.M. (1981) *Extrusion of Foods*, CRS Press, Florida.

5 Mercier, C., Linko, P., and Harper, J.M. (1989) *Extrusion Cooking*, American Association of Cereal Chemists, Inc., St. Paul, Minnesota, USA.

6 Mościcki, L. and Pyś, D. (1993) Bębny uszlachetniające ekstrudaty typu 01IZ1, Postępy Techniki Przetwórstwa Spożywczego (in Polish). 2, 31–33.

7 Bruin, S., van Zuilichem, D.J., and Stolp, W. (1978) A review of fundamental and engineering aspects of extrusion of biopolymers in a single screw extruder. *J. Food Process Eng.*, **2**, 1–37.

8 Mościcki, L. (2003) Effect of screw configuration on quality and SME value of corn extrudate. Teka Commission of Motorization Power Industry in Agriculture, vol. III, pp. 182–186.

9 van Zuilichem, D.J., Lamers, G., and Stolp, W. (1975) Influence of process variables on quality of extruded maize grits. Proceedings of the 6th European Svmposium on Engineering and Food Quality, Cambridge.

10 Bagley, E.B. (1957) End corrections in the capillary flow of polyethylene. *J Appl. Phys.*, **28**, 624.

11 Metzner, A.B. (1959) Flow behaviour of thermoplastics, in *Processing of Thermoplastic Materials* (ed. E.C. Bernhardt), Van Nostrand-Reinhold, New York.

12 van Zuilichem, D.J., Bruin, S., Janssen, L.P.B.M., and Stolp., W. (1980) Single

screw extrusion of starch and protein rich materials, in *Food Process Engineering Vol. 1: Food Processing Systems* (eds P. Linko, V. Malkki, J. Oikku, and J. Larinkariels), Applied science, London, pp. 745–756.

13 Janssen, L.P.B.M., Spoor, M.W., Hollander, R., and Smith, J.M. (1979) Residence time distribution in a plasticating twin screw extruder. *AlChE*, **75**, 345–351.

14 Juśko, S., Mitrus, M., Mościcki, L., Rejak, A., and Wójtowicz, A. (2001) Wpływ geometrii układu plastyfikującego na przebieg procesu ekstruzji surowców roślinnych (in Polish). *Inżynieria Rolnicza*, **2**, 124–129.

15 Janssen, L.P.B.M. (1978) *Twin Screw Extrusion*, Elsevier Scientific Publishing Company, Amsterdam.

16 Janssen, L.P.B.M., Noomen, G.H., and Smith, J.M. (1975) The temperature distribution across a single screw extruder channel. *Plast. Polym.*, **43**, 135–140.

17 Janssen, L.P.B.M. and Smith, J.M. (1979) A comparison between single and twin screw extruders. Proceedings of the 1st. Conference on Polymer Extrusion. Plastics and Rubber Institute, London, June, pp. 91–99.

18 Mościcki, L. (1986) Wpływ warunków ekstruzji na przebieg i efektywność procesu ekstruzji surowców roślinnych (in Polish). *Roczniki Nauk Rolniczych*, **76-C-2**, 223–232.

19 Pinto, G. and Tadmor, Z. (1970) Mixing and residence time distribution in melt screw extruders. *Polvm. Eng. Sci.*, **10**, 279–288.

20 Rosse, J.L. and Miller, R.C. (1973) Food extrusion. *Food Technol.*, **27** (8), 46–53.

21 van Zuilichem, D.J., Buisman, G., and Stolp, W. (1974) Shear behaviour of extruded maize. IUFoST Conference, Madrid, September, pp. 29–32.

2
Engineering Aspects of Extrusion

Dick J. van Zuilichem, Leon P.B.M. Janssen, and Leszek Mościcki

2.1
Mass Flow and Temperature Distribution in a Single-Screw Extruder

There are two main categories of factors that play a key role during the extrusion-cooking process: those related to the type of processed material and those derived from the operational and technical characteristics of the extruder.

The most important factors in the first category are: moisture, viscosity and chemical composition of the raw material; in the second category they are: the compression ratio and configuration of the screw, its rotational speed, process temperature and the pressure range applied in the barrel. The range of adjustable factors includes flow intensity, process temperature and the size of the die openings. All these factors are in close correlation with each other and have a decisive impact on the quality of the extruded product. Their mutual relations are the subject of several theoretical deliberations. This is true especially for vegetable material where, for example, the presence of Newtonian flows is limited due to the large variety of rheological characteristics of the raw material.

2.1.1
The Theory of Mass Flow and Temperature Distribution

To be able to comprehend the mechanism of mixing, flow rate, pressure distribution, and so on, it is necessary to determine the type of flow in the extruder barrel. The basic problem in describing flow patterns inside the extrusion-cooker is that the flows in the compression and metering section are non-Newtonian and non-isothermal.

The most important simplifications are the assumptions of steady state, negligible inertia and gravity forces and fully developed incompressible fluid flow. With these assumptions the flow of material in the barrel of the extruder is reduced to a movement in a slot having width w and height h and one wall moving with a speed of πND against a gradient pressure $\partial p/\partial z$ (Figure 2.1). The flow of the material is analyzed in relation to the rectangular coordinate system x, y, z, which rotates at the angular velocity around the system axis linked with the axis of the extruder barrel.

Extrusion-Cooking Techniques: Applications, Theory and Sustainability. Edited by Leszek Moscicki

ISBN: 978-3-527-32888-8

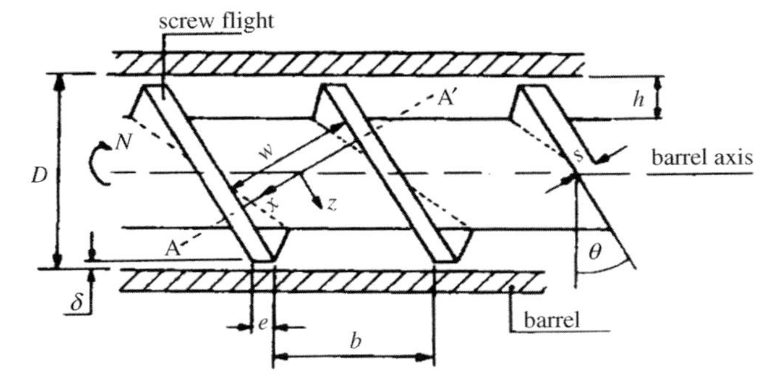

1. Section of extruder with definition of geometry

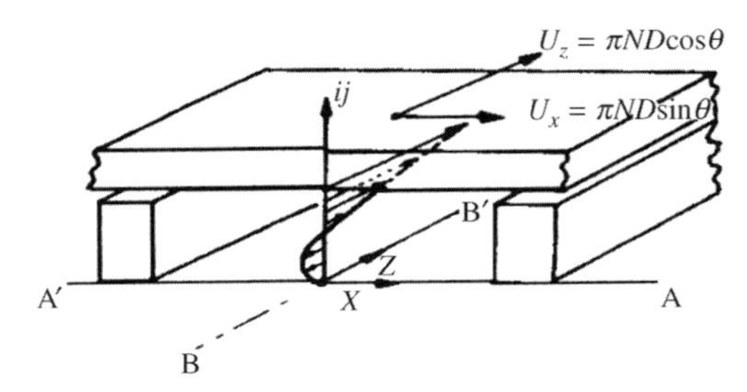

2. Simplified flow geometry (section AA′)

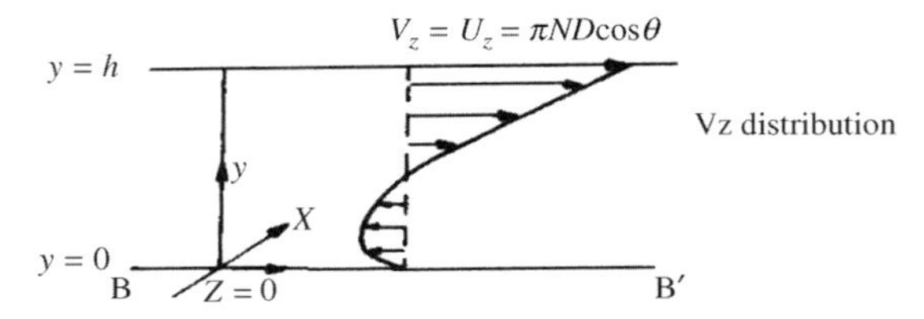

3. Flow profile in coordinate system used (section BB′)

Figure 2.1 A graphic interpretation of the theory of material movement in a single-screw extruder [1].

The velocity of the barrel wall can now be divided into two components: in the cross channel direction $U_x = \pi ND\sin\theta$ and in the channel direction $U_z = \pi ND\cos\theta$ (where N is the screw rotational speed, D the internal diameter of the extruder barrel, and θ the angle of screw flight). The flow in the z-direction is the result of two forces: the drag caused by the z component of the relative velocity of the barrel (U_z) and the pressure gradient in the direction of z, since the value of pressure p gradually grows in this direction [32, 33].

The simplest case arises when the material behaves as a Newtonian fluid with temperature-independent viscosity and when the velocity components in the y-direction near the flights are neglected. This assumption is a reasonable approximation for shallow channels ($h/w < 0.1$).

Theoretical calculations of the distribution of flow speeds V_x and V_z, and the calculation of temperature distribution during extrusion are given in more extensive studies discussing the fundamental dependences, formulas and ways of calculating the above-mentioned values [1–5]. The formula for the calculation of speeds V_x and V_z is presented in Section 2.3, in Figure 2.3.

The material flow rate Q_v can be determined from the component of velocity V_z by integration over the entire section of the inter screw flight space and by multiplying the result by the number of screw flights, k:

$$Q_v = k \int_0^w \int_0^h V_z dx dy \tag{2.1}$$

Inserting the velocity component defined according to Figure 2.3 gives:

$$Q_v = \frac{1}{2} hkwU_z \left(1 - \frac{1}{3} P\right) = \frac{1}{2} hkwU_z - \frac{1}{12} \frac{h^3 kw}{\mu} \left(\frac{\partial p}{\partial z}\right) \tag{2.2}$$

where: μ is the Newtonian coefficient of material viscosity and:

$$P = \frac{h^2}{2\mu U_z} \left(\frac{\partial p}{\partial z}\right) \tag{2.3}$$

Griffith [2] identified Q_v for the material whose viscosity depends on temperature. The conditions in which the temperature is fully dependent on the coordinates x and z are relatively rare. These solutions can be considered valid if they are used for general calculations, that is, much simplified.

As shown above in the formula for Q_v, no material flow has been taken into account between the flights of the screw and the walls of the barrel. If Equation 2.1 is supplemented by this third element of flow, then we get:

$$Q_v = \frac{1}{2} hkwU_z - \frac{1}{12} \frac{h^3 kw(\partial p)}{\mu(\partial z)} - \frac{kwU_x \delta}{2tg\theta} \tag{2.4}$$

where δ is the distance between the barrel wall and the screw flight.

The processed material flows through the die. Assuming that the die opening is in the shape of a cylinder with diameter d and length l, then Q_v can be written as:

$$Q_V = \frac{\pi d^4 \Delta p}{128 \mu l} \tag{2.5}$$

Assuming that the pressure in the die opening is identical to that at the end of the barrel, then replacing $\partial p / \partial z$ by $\Delta p / (L/\sin\theta)$ in Equation 2.4, where L is the length of the screw, it will be possible to exclude Δp in Equation 2.5. Then the flow rate of extruded material can be shown as a non-dimensional equation [1, 3, 6]:

$$\frac{Q_v \sin\theta}{kwhNL} = \frac{\pi}{2} \frac{\left(1 - \frac{\delta}{h}\right)\left(\frac{D}{L}\right) \sin\theta \cos\theta}{1 + \frac{32}{3\pi}\left(\frac{h}{d}\right)^3 \frac{l}{d} \frac{wk}{L} \sin\theta} \equiv \Lambda \tag{2.6}$$

As mentioned at the outset, during the extrusion-cooking of vegetable materials, the processed material often behaves like a non-Newtonian liquid. The main reason for this is the volatility of the rheological features of the processed material which, in turn, depend on the conditions of material compression and the scope of physical and chemical transformations within the process. In the available literature, one can indeed identify some information about these features but it is incomplete and relates only to some selected vegetable materials. Hence, only after having collected the appropriate data, will it be possible to propose a flow model for the material in question.

2.1.2
Residence Time Distribution of the Material in the Extruder

When describing the extrusion process, it is of great importance to define the residence time for material particles in the extruder. On the basis of this time distribution, it is possible to establish the degree of mixing of the material, anticipate the course of plasticization as well as the extent and degree of uniformness in the deformation of the stream of liquid material during extrusion. Residence time is largely the result of the distribution of the velocities inside the device and the length of the screw. Although it is possible to calculate the residence time distribution for particular zones in the extruder from the flow velocities, practice shows that it is empirical evidence that provides the best results.

The theory of the distribution of mixing time has been developed by process engineering [7]. Many authors have explored the characteristics of flow and mixing dynamics in extrusion processes, in particular the extrusion of plastics [3, 4, 6, 8–10].

The average residence time of the material in a fully filled extruder can be calculated from the ratio between the volume and the volumetric throughput:

$$\bar{t} = \frac{kwhL}{Q_v \sin\theta} \tag{2.7}$$

Because the throughput is proportional to the rotation rate (at constant end pressure) the average residence time for particles in the extruder is proportional to the inverse of the rotational speed of the screw N.

Particular attention should be given to the proper study of the function of time distribution $E(t)$ and $F(t)$.

$E(t)$ is called the internal age distribution and can be defined as the fraction of the material that flows through the die opening after a residence time between t and $t + \mathrm{d}t$. The function $F(t)$ is defined by integration of $E(t)$, being the cumulative function of the internal age distribution:.

$$F(t) = \int_0^t E(t)\mathrm{d}t \tag{2.8}$$

The function $(1 - F(t))$ is called the exit age distribution and is particularly important in food extrusion, because it represents the fraction of material that remains longer than a given time t in the extruder. The larger this fraction, the greater chance that the product will show signs of burning.

The simplest case, for which the average residence time of material elements in the extruder can be determined, is the flow of Newtonian fluids. Pinto and Tadmor [6] examining the flow of material in a single-screw extruder and assuming Newtonian flow and a constant material viscosity μ defined the average residence time as the first moment of the exit age distribution:

$$\bar{t} = \int_{0}^{\infty} tE(t)\mathrm{d}t \tag{2.9}$$

For a single-screw food extruder $\bar{t} = {}^4/_3 t_0$, where t_0 is the minimum residence time for the particles inside the extruder or the break through time. This residence time occurs for fluid elements that remain in the channel at the level of $y/h = 2/3$.

The literature offers relatively limited data on the extrusion time of vegetable materials. Zuilichem *et al.* [9] using the technique of radioactivity studied residence time distribution for maize in a single-screw extrusion-cooker. As a result of multiple trials they found that in the case of maize the residence time for its particles inside that type of extruder ranged from 30 to 45 s.

2.2 Energy Balance

It is a widespread opinion that the extrusion technique is highly energy consuming. However, this is not justified in the light of research, especially in relation to most conventional methods of heat treatment of vegetable material. Let us have a closer look at the basic issues of energy consumption of extrusion-cooking processes.

The rotation of the extruder screw provides input of mechanical energy. Through viscous dissipation, most of this energy is converted to heat in the processed material but some goes to increase the pressure in the material and its kinetic energy. The total energy balance $\mathrm{d}E$ consists of the following components:

$$\mathrm{d}E = \mathrm{d}E_H + \mathrm{d}E_p + \mathrm{d}E_k + \mathrm{d}E_\delta \tag{2.10}$$

Where: $\mathrm{d}E_H$ is the viscous energy dissipation in the channel, $\mathrm{d}E_p$ the energy consumed to increase the pressure of the processed material, $\mathrm{d}E_k$ the energy consumed to increase kinetic energy, $\mathrm{d}E_\delta$ the increase in heat energy in the area between the screw and the cylinder. Each of the above items will be discussed separately with the assumption that the velocities of mass flow in the extruder are small, and consequently $\mathrm{d}E_k$ is negligible.

2.2.1 Components of Energy Balance

In a channel element between screw flights, with dimensions $wh\mathrm{d}z$, the energy condition may be expressed by the following equation [5, 11]:

$$dE_H + dE_p = k\mu(\pi ND)^2 \frac{w}{h}(\cos^2\theta + 4\sin^2\theta)dz + k\pi ND\frac{wh}{2}\frac{\partial p}{\partial z}\cos\theta dz \qquad (2.11)$$

The dissipation of heat energy in this channel element is proportional to μ and N^2. The energy requirements for the pressure increase are directly proportional to N, which is related to Q and the pressure increase along the channel. In most cases, the energy to increase the pressure of the processed material is very small compared to the viscous dissipation term due to the relatively low Δp and high μ involved.

dE_δ in an area of the flight clearance due to drag flow can be expressed by the following equation:

$$dE_\delta = k\left[\int_0^e \tau V dx\right] dz \qquad (2.12)$$

where e is the screw flight tip width and the shear stress τ can safely be approximated by:

$$\tau = \mu_\delta \frac{\pi ND}{\delta} \qquad (2.13)$$

The viscosity in the clearance is μ_δ to denote the fact that it may be significantly different from the viscosity in the channel because of temperature and shear effects.

Using $V = \pi DN$ and integrating gives:

$$dE_\delta = k\mu_\delta \frac{(\pi DN)^2 e}{\delta} dz \qquad (2.14)$$

This equation shows the high significance of the clearance between the screw and the barrel surface. The smaller the gap the higher the power needed to the drive the extruder.

2.2.2 Total Power Input to a Screw

Starting with the total Equation 2.10 for power input, the following equation emerges:

$$dE = k\left[\mu(\pi ND)^2 \frac{w}{h}(\cos^2\theta + 4\sin^2\theta) + \pi NDw\frac{h}{2}\frac{\partial p}{\partial z}\cos\theta + \mu_\delta(\pi ND)^2\frac{e}{\delta}\right]dz \qquad (2.15)$$

Assuming that:

$$dL = dz\sin\theta \qquad (2.16)$$

after integration we obtain the total power input as:

$$E = k\frac{(\pi ND)^2 L}{\sin\theta}\left[\mu\frac{w}{h}(\cos^2\theta + 4\sin^2\theta) + \mu_\delta\frac{e}{\delta}\right] + k\frac{\pi NDwh}{2}\Delta p\cos\theta \qquad (2.17)$$

This equation is sometimes rewritten as:

$$E = G_2 N^2 \left[\mu \frac{w}{h}(\cos^2\theta + 4\sin^2\theta) + \mu_\delta \frac{e}{\delta}\right] + G_1 N \Delta p \qquad (2.18)$$

where:

$$G_2 = \frac{k(\pi D)^2 L}{\sin\theta} \qquad (2.19)$$

and

$$G_1 = \frac{\pi}{2} D^2 h\left(1 - \frac{ek}{\pi D \sin\theta}\right)\sin\theta\cos\theta \qquad (2.20)$$

In extrusion theory we often make use of the specific mechanical energy (SME), that is, the energy consumption per unit of mass of processed material. In many food extrusion processes where the pressure is low the flow rate is almost equal to the drag flow or G_1N. In this case it is expected that the SME will increase linearly with the speed of the screw N for Newtonian fluids at a constant temperature T. For pseudoplastic materials, the viscosity is proportional to the shear rate $\dot{\gamma}$ which is proportional to N. Therefore the viscosity is proportional to N^{n-1} and the SME can be expected to increase as N^n.

Strictly speaking, Equations 2.17 and 2.18 describe the power input for an extruder operating with Newtonian fluids only. These equations can be used to approximate the energy balance for non-Newtonian fluids if the apparent viscosity η is substituted for the Newtonian viscosity μ. In this case, η needs to be evaluated for the shear rate $\dot{\gamma}_H$ and temperature in the channel. Although the shear rate varies with the location in the channel, for low-pressure processes the average shear rate can be approximated by:

$$\dot{\gamma}_H = \frac{\pi D N}{h} \qquad (2.21)$$

Similarly, the viscosity in the clearance between the screw tip and the barrel wall needs to be evaluated at $\dot{\gamma} = \pi ND/\sigma$ and the temperature in the clearance.

Bruin *et al.* [1] divided both parts of Equation 2.15 by $\mu k N^2 D^3$ and obtained a power number similar to that used in the analysis and design of mixing systems in the viscous flow region. It should be noted that the power number is a constant for a given extruder and die design. In the literature there are insufficient examples of practical verification of the energy consumption models during extrusion. However, these models give a useful approximation for solving problems with baro-thermal treatment of biopolymers.

The application of Equation 2.18 requires a thorough knowledge of the rheology of food materials as a function of process temperature. Most of them are non-Newtonian fluids, therefore it is important to know the shear ratio $\dot{\gamma}$.

Harmann and Harper [12] calculated the torque requirements for extrusion-cooking a rehydrated pregelatinized corn dough and compared the results with actual measurements. Their experiment was conducted in isothermal conditions with $\dot{\gamma}$ calculated according to Equation 2.21. The values of torque obtained by dividing Equation 2.18 by N were 17 to 73% higher than the values actually measured. They assigned the inability to predict the current values in this equation to insufficient data on the material temperature at the surface of the barrel and the filling degree of the screw. The equation might have over predicted the theoretical power demand, because the temperature at the surface of the barrel was probably higher than the temperature of the entire material, with a small degree of filling of the screw.

According to the investigations of Moscicki *et al.* [8, 11] the demand for energy during extrusion of vegetable material decreases with increase in the process temperature and moisture of the processed material. Generally speaking, it can be assumed that good results are obtained at an energy consumption (SME) with an average of 0.1 to 0.2 $kWh\,kg^{-1}$.

The impact of an increase in speed of mass flow on energy consumption during extrusion-cooking has been investigated by Tsao [13], who used 14 different screws using a high moisture food dough. His results showed specific power increasing approximately as N^n within a range of a threefold increase in screw speed, as expected. Actual power requirements were less than those calculated and could be attributed to the actual dough temperature being higher than that measured at the barrel surface.

A number of SME measurements for cereal–soybean mixtures for a simple autogenous extruder at a moisture content of 12 to 20% have been made by Harper [14]. These data show ranges in SME of 0.08 to 0.16 $kWh\,kg^{-1}$. He also confirmed that energy consumption did not increase in proportion to N as anticipated. There is an urgent need for a more accurate understanding of the effects of moisture and shear rate on viscosity and the correct measurement of material temperature.

Bruin *et al.* [1] examined energy consumption during the extrusion of cereals at different moisture contents in a single-screw extrusion-cooker fitted with different dies and screws at a single screw speed. Their data showed not only a reduction in SME with increased moisture, but also an increase at higher moisture. They attributed these results to increased gelatinization at higher moistures.

Above we have analyzed the energy of the extruder's screw drive related to the transformation of mechanical energy into heat in the channel and flight clearance. During extrusion-cooking, this mechanical energy is principally dissipated in the form of heat, which causes temperature, chemical and perhaps phase changes in the processed material. The amount of energy dissipated may be insufficient in relation to the thermal requirements, so very often heat is added with a heating system for the barrel and sometimes for the screw. Direct addition of steam to the processed material is also used in order to raise its temperature and hydration.

The energy balance per unit of time can be expressed as follows:

$$\frac{E_t}{\Delta t} = \frac{E}{\Delta t} + q \tag{2.22}$$

and

$$\frac{E_t}{\Delta t} = Q_v \left[\int_{T_1}^{T_2} c_p \mathrm{d}T + \int_{p_1}^{p_2} \frac{\mathrm{d}p}{\varrho} + \Delta H^\circ + \Delta H_{s1} \right] \quad (2.23)$$

where E_t is the total net energy added to the extruder, E the mechanical energy dissipated (from Equation 2.6), Δt the process time, q the heat flux to dough (+) or loss (−), c_p the specific heat, ΔH° the heat of reaction/unit mass dough, ΔH_{s1} the latent heat of fusion/unit mass of dough, ϱ the density of the material, 1 – material feeder hole, 2 just behind the die.

A certain amount of heat is also generated as a result of chemical reactions such as starch gelatinization, protein denaturation, browning, and so on. Sahagun and Harper, [15] using blends of 70% maize and 30% soybean, found that approximately 15% of total energy added to the autogenous extruder was unaccounted for by sensible heat changes, pressure increase and heat loss. These differences are attributed to the heat of reaction associated with chemical changes occurring in food materials and heat losses to the surroundings (radiation and convection).

The latent heat of fusion is also included in the energy balance to account for a relatively small amount of energy associated with the melting of solid lipids materials that may be part of a food formulation. To prepare a comprehensive model of energy balance in the extrusion process still requires much work and many experiments.

2.3
Mass and Heat Transfer in a Twin-Screw Extruder

There are two main types of food extruders: single- and twin-screw. The second type is divided into the so-called co-rotating and counter-rotating extruders. Counter-rotating twin-screw extruders work as positive-displacement pumps with closed C-shaped chambers between the screws [4] – this minimizes mixing but also backflow caused by the increase in pressure. In co-rotating extruders, material is transported steadily from one screw to another. The flow mechanism can be described by a combination of drag flow and a positive displacement caused by the pushing action of the screw-set in the intermeshing region. Co-rotating extruders usually work at a higher screw speed ranging from 300 to 600 rpm.

Although currently designed screws have a modular structure of different geometry, they are generally divided according to three different screw sections:

- the feed section, which ensures that sufficient solid material is transported into the screw;
- the compression section, in which the material is heated and processed into a dough-like mass,
- the metering section, in which the screw configuration feeds the die constantly with material.

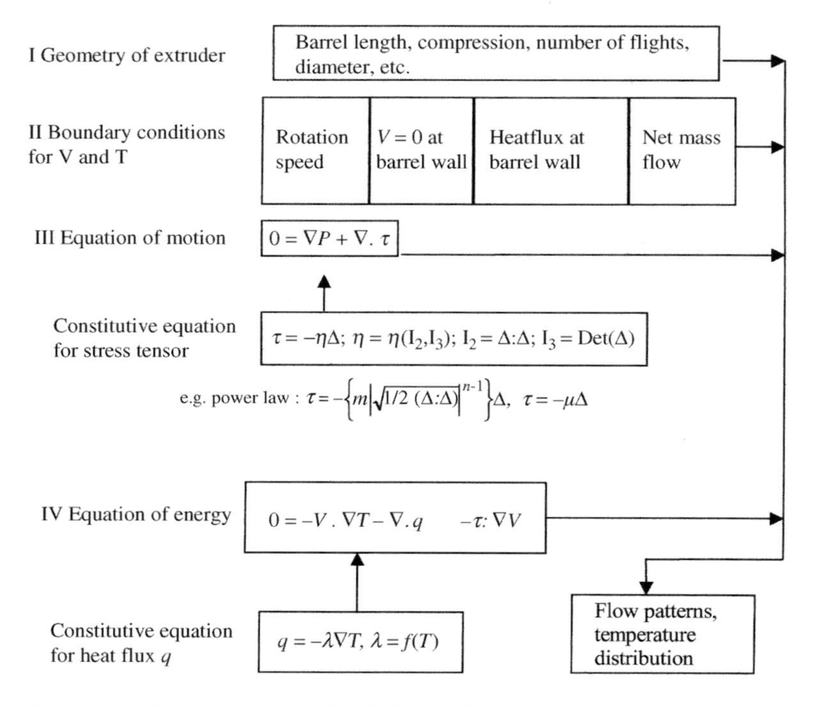

Figure 2.2 Schematic approach for calculation of flow patterns in an extruder [1].

The basic problem in the description of the model of flow inside the extruder is that the flow in the compression section and in the metering section is non-Newtonian and not isothermal. Therefore, the equation of movement and the equation of energy, should be presented so as to obtain a realistic model of flow. The main procedure is shown in the diagram in Figure 2.2. The set of equations can be solved only with a few approximations.

The most important are the assumptions of steady states, negligible inertia and gravity forces and of fully developed incompressible fluid flow. Other assumptions are set out in Section 2.1. With these assumptions and simplifications the solution of the equation of motion and the energy equation is possible. The most simple case arises when the material behaves as a Newtonian fluid with temperature-independent viscosity and when the velocity component in the y-direction near the flights is neglected. This second assumption is a reasonable approximation for shallow channels ($h/w < 0.1$).

Figure 2.3 illustrates how velocity profiles could be derived and, successively, the heat flux at the barrel surface and the screw root can be calculated. It is interesting to note that the equation of energy is fairly difficult to solve even for this simple rheological behavior because of the convection term. Only when adiabatic extrusion is assumed and the temperature gradient in the y-direction is assumed to be negligible, can a relatively simple result be obtained.

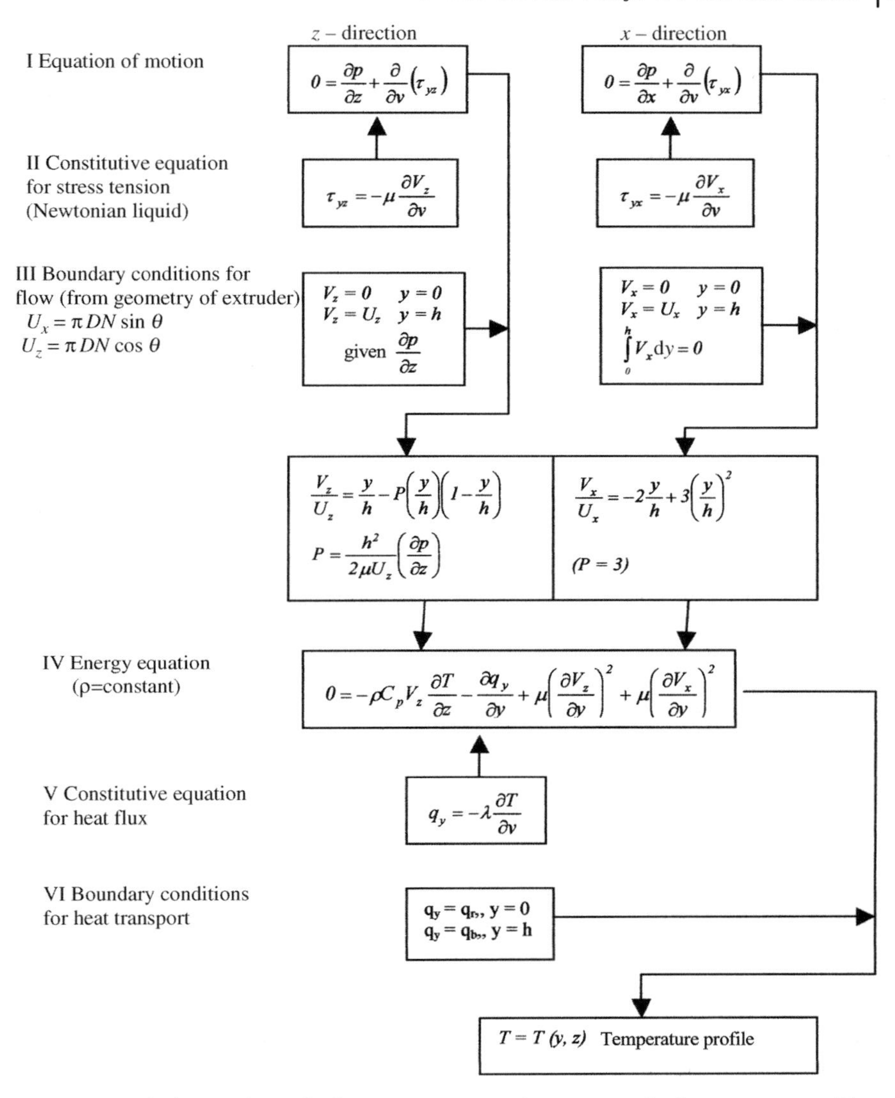

Figure 2.3 Calculation scheme for flow patterns in a single screw extruder for an incompressible Newtonian fluid with constant heat conductivity and viscosity [1].

2.3.1 Heat Transfer

The main problem in modeling heat transfer in food extruders is the inclusion of all variables in a complete analysis. This seems to be quite impossible because of the computational effort involved although in many cases some of the variables can be ignored as unimportant for the practical set of conditions under consideration. This means that it is possible to make some assumptions and to derive a solution which is useful under certain operating conditions. Care has to be taken, however, that the model is not used for an unjustified extrapolation to other circumstances.

As commonly known, most food materials behave like non-Newtonian fluids. Griffith [16], Pearson [17], Yankov [18] and Yacu [19] developed models which give a better understanding of the extrusion of such materials. Griffith [16] used a numerical analysis to solve differential equations for fully developed velocity and temperature profiles for single screw extrusion of power-law fluids. Pearson [17] solved the equations of motion and energy for transverse channel flow and a superimposed temperature profile for power-law fluids. Yankov [18] proposed equations for movement of non-Newtonian fluids, assuming that the temperature did not change along the channel. Unfortunately, these models are doubtful because of the assumptions used in the description of the phenomenon, especially when it comes to extrusion of food. The most thorough description of the non-Newtonian behavior of mass was proposed by Yacu [19], which has been taken into account in developing the model proposed by Zuilichem [5, 20, 21].

The first examination of two-phase flow in the melting zone in a single-screw extruder was carried out by Maddok [22]. He concluded that solid particles in contact with the hot surface of the barrel partially melt and smear a film of molten polymer over the barrel surface. This film, and probably some particles, are dragged partly along the barrel surface, and when they meet the advancing flight, they are mixed with previously molten material. The molten material collects at an area in the pushing flight (meltpool), whereas the forward portion of the channel is filled with solid particles. The width of a solid bed, X, gradually decreases toward the outlet of the screw. The melting process ends when the solid bed disappears.

Based on the observations of Maddock [22], Tadmor and Klein [23] proposed a model which provided the equations that are required to calculate the length of the melting zone. This length depends on the physical properties of the polymer and the operating conditions. The only melting mechanism takes place between the hot barrel and the solid bed. The molten mass is transported by the movement of the barrel relative to the solid bed, until it reaches a screw flight. The leading edge of the advancing flight scrapes the melt of the barrel surface and forces it into a meltpool. This meltpool needs space and moves into the solid bed; the solid bed deforms and its width decreases. It is assumed that the melt film thickness profile does not change in the melting zone and that the melt film thickness equals the flight clearance. Shapiro [24] and Vermeulen *et al.* [25] concluded that the melt film thickness cannot be constant and will increase. This decreases the heat flux to the solid bed and, consequently, increases the melting length.

Over the years Tadmors model has been improved and adjusted by himself and many other authors, among others, by Dekker [7] and Lindt [26].

For the heat transfer coefficient two different mechanisms have to be distinguished, heat transfer at the barrel and heat transfer at the screw. At the barrel the flight of the rotating screw wipes a certain area on the inner barrel surface, a fresh layer of polymer becomes attached to the same region and remains there for, approximately, one revolution. The amount of heat penetrating into this layer during that time by pure conductivity heat transfer is then removed with the polymer layer and is thought to be homogeneously distributed throughout the bulk of the polymer in the screw channel [27]. This process of heat penetration during extrusion can often be considered as a non-stationary penetration in a semi-infinite medium. The heat balance is given by the following equation:

$$\varrho c_p \frac{\partial T}{\partial t} = \lambda\left(\frac{\partial^2 T}{\partial x^2} + \frac{\partial^2 T}{\partial y^2} + \frac{\partial^2 T}{\partial z^2}\right) \tag{2.24}$$

or in one dimension:

$$\frac{\mathrm{d}T}{\mathrm{d}t} = a\frac{\mathrm{d}^2 T}{\mathrm{d}y^2} \tag{2.25}$$

Where λ is the thermal conductivity, and a the thermal diffusivity. The solution of this differential equation gives the temperature distribution in the thermal boundary layer and can be written as:

$$T - T_0 = 1 - \mathrm{erf}\left[\frac{x}{\sqrt{\pi a t}}\right]$$

From this expression the time average heat transfer coefficient at the barrel wall can be calculated to be [4]:

$$\langle \alpha \rangle = \sqrt{\frac{4}{\pi}\lambda \varrho c_p m N}$$

Where ϱ and c_p are the density and specific heat of the dough and m and N are the number of flights starts and the rotation rate of the screws.

In contrast to the heat transfer with surface renewal that occurs at the barrel wall the heat transfer at the screw surface has to be described as heat transfer to a semi-infinite flowing medium, for which the energy balance can be approximated by:

$$V_x \frac{\partial T}{\partial x} = a\frac{\partial^2 T}{\partial y^2} \tag{2.26}$$

with boundary conditions:

$$\begin{array}{rcl} x < 0 & & T = T_o \\ x > 0 & x \to \infty & T = T_o \\ x > 0 & y = 0 & T = T_w \end{array}$$

with the assumption that:

$$V_x = \dot{\gamma} y \tag{2.27}$$

Where: $y = 0$ indicates the location on the barrel wall at uniform temperature T_w.

The temperature profile in the thermal boundary layer is approximated by a parabolic function:

$$\frac{T - T_o}{T_w - T_o} = \left(1 - \frac{y}{\delta}\right)^2 \tag{2.28}$$

where δ is the (unknown) boundary layer thickness. After combining this with Equation 2.26 and integrating, the boundary layer thickness is given by:

$$\delta = \left(\frac{36ax}{\dot{\gamma}}\right)^{\frac{1}{3}} \tag{2.29}$$

And for the heat flux we can write:

$$\Phi''_{\mathrm{w}} = 2\lambda\left(\frac{36ax}{\dot{\gamma}}\right)^{-\frac{1}{3}}(T_{\mathrm{w}}-T_{\mathrm{o}}) \tag{2.30}$$

Therefore the average heat transfer coefficient at the screw on location x can be approximated by:

$$\alpha_x = 2\lambda\left(\frac{36ax}{\dot{\gamma}}\right)^{-\frac{1}{3}} \tag{2.31}$$

Inspection of the two different heat transfer coefficients shows that, in general food extrusion processes, the heat transfer at the wall is significantly more important than that at the screw. However, an important exception has to be made. In the derivation of the expression for the heat transfer coefficient at the barrel wall it is wiped clean completely. In practice, where there exists a gap between the screw flight and the barrel wall a small layer of material will remain at the wall, forming an insulating layer that can decrease the heat transfer considerably. Solving Equation 2.26 for this case leads to a correction for the heat transfer coefficient at the wall:

$$\langle\alpha\rangle = \sqrt{\frac{4}{\pi}\lambda\varrho c_p mN} \times \exp\left(\frac{\delta}{2}\sqrt{\frac{mN}{a}}\right)$$

where δ stands for the clearance between screw flight and barrel wall [28].

2.3.2 Model by Yacu

Yacu [19] divided the twin-screw extrusion-cooker into three main sections: solid conveying zone, melt pumping zone and melt shearing zone. The temperature and pressure profiles are predicted for each section separately. The following assumptions were made:

- The rheology of the molten material is described by a non-Newtonian, non-isothermal viscosity model taking into account the effect of moisture and fat content. The extruder is operating under steady state conditions.
- Steady state behavior and uniform conditions exist.
- The melt flow is highly viscous and in the laminar flow regime.
- Gravity effects are negligible and ignored.
- The screw is assumed to be adiabatic.
- The degree of cooking is assumed to be uniform over the cross-section.

2.3.2.1 **Solid Conveying Section**

Twin-screw food extruders operate mostly under starved feed conditions and the throughput is determined by the feeding unit, and the extruder's screw speed and torque. Therefore the screws in the feeding section are partially filled and no pressure is developed. The dispersion of mechanical energy is negligible because of these conditions. Heat is transferred by conduction from the barrel to the food material.

To simplify the analysis and taking into account that mixing within the channel is reasonably good in a co-rotating extruder, the heat transfer is assumed to be controlled only by convection.

Constructing a heat balance across an element normal to the axial direction and solving for $T$with the boundary condition at $x = 0$ and $T = T_f$, results in the equations:

$$Q_m c_{ps} T + F U_s A(T_b - T)dx = Q_m c_{ps}(T + dT) \tag{2.32}$$

which gives:

$$T = T_b - (T_b - T_f)e^{\frac{-FU_s Ax}{Q_m c_{ps}}} \tag{2.33}$$

where Q_m is the feed rate, c_{ps} the heat capacity of the solid matter, F the degree of fill, A the surface area, T_b the temperature of the barrel, T_f the temperature of the feed material, and U_s the pseudo heat transfer coefficient.

The heat transfer coefficient U_s to the powder in the feed zone, for practical reasons estimated as the heat transfer coefficient based on the penetration theory, would give unrealistic values. Phase discontinuity and the existence of additional resistance to heat transfer between the solid particles are the main reasons for this.

2.3.2.2 **Melt Pumping Section**

In this section the change of the material from solid food-powder to a fluid melt is assumed to take place abruptly. Martelli [29] assumed that the energy was dissipated within the channel and due to the leakage flows in the various gaps. He defined four locations where energy was converted:

- in the channel:

$$Z_1 = \frac{\pi^4 D_e^3 D tg\theta}{2h} \mu N^2 \tag{2.34}$$

where h is the channel depth and D_e the equivalent twin-screw diameter.

- between the flight tip of the screw and the inner surface of the barrel:

$$Z_2 = \frac{\pi^2 D^2 e C_e}{\delta} k\mu N^2 \tag{2.35}$$

where C_e is the equivalent twin-screw circumference

- between the flight tip of one screw and the bottom of the channel of the other screw:

$$Z_3 = \frac{8\pi^2 I^3 e}{\varepsilon} k\mu N^2 \tag{2.36}$$

Where ε is the clearance between the flight tip channel bottom of two opposite screws, and I is the distance between the screw shafts.

- between the flights of opposite screws parallel to each other:

$$Z_4 = \frac{\pi^2 I^2 h\sqrt{D^2 - I^2}}{2\sigma} k\mu N^2 \tag{2.37}$$

where σ is the clearance between flights of opposite screws parallel to each other.

The total energy converted per channel per screw turn was therefore expressed as:

$$Z_p = Z_1 + Z_2 + Z_3 + Z_4 = C_{1p}\mu_p N^2 \tag{2.38}$$

Where: μ_p is the viscosity of the product; C_{1p} can be defined as the pumping section screw geometry-factor and is described by:

$$C_{1p} = \frac{\pi^4 D_e^3 Dtg\theta}{2h} + k\left(\frac{\pi^2 De^2 C_e}{\delta} + \frac{8\pi^2 I^3 e}{\varepsilon}\frac{\pi^2 I^2 h\sqrt{D^2 - I^2}}{2\sigma}\right) \tag{2.39}$$

Because the length of the screw channel per screw turn equals $\pi Dtg\theta$, the average amount of heat generated within an element of thickness dx can therefore be evaluated as:

$$dZ_p = C_{1p}\mu_p N \frac{dx}{\pi Dtg\theta} \tag{2.40}$$

The overall shear rate on the product, while passing through the pumping zone, including the amount taken up by leakage flows, can be estimated as:

$$\dot{\gamma} = N\frac{\sqrt{C_{1p}}}{V_p} \tag{2.41}$$

The viscosity μ_p is dependent on shear rate ($\dot{\gamma}$), temperature (T), moisture content (MC) and fat content (FC):

$$\mu_p = \mu\dot{\gamma}^{-n_1} e^{-a_1(MC - MC_o)} e^{-a_2 FC} e^{-b_1 \Delta T} \tag{2.42}$$

where a_1 is the moisture coefficient of viscosity, a_2 the fat coefficient of viscosity, b_1 the temperature coefficient of viscosity, and n_1 the power-law index.

This rheological model was developed by Yacu [19] for a wheat starch. The various indices are determined by multiple regression analysis to fit the rheological model.

Constructing a heat balance across an element in the melt pumping zone, coupled with the boundary conditions: $T = T_m$ at $x = X_m$, gives the relation:

$$Q_m c_{pm} T + \alpha A(T_b - T)dx + \frac{C_{1p}\mu_p N^2 e^{-b_1 T}}{\pi Dtg\theta} dx = Q_m c_{pm}(T + dT) \tag{2.43}$$

or

$$\frac{dT}{dx} = C_{2p} e^{-b_1 T} + C_{3p}(T_b - T) \tag{2.44}$$

where:

$$C_{2p} = \frac{C_{1p}\mu_p N^2}{\pi D tg\theta Q_m c_{pm}} \quad (2.45)$$

and

$$C_{3p} = \frac{FA}{Q_m c_{pm}} \quad (2.46)$$

In these formulas c_{pm} is the heat capacity of the molten material, α the heat transfer coefficient, and T_m the temperature of the molten material.

One of the assumptions made by Yacu [19] was that the moment the mass enters the melt pumping section the screws become completely filled. In this calculation the mass is subjected to the total shear stress. However, most twin-screw extruders are starved fed which results in a partially low filling degree. It would, therefore, be reasonable to conclude that the shear the product receives is lower than Yacu assumes.

2.3.3
Model by van Zuilichem

The unique feature of the model developed by van Zuilichem is that it calculates the total transferred heat in the extruder for every position along the screw axis. The model is composed of two major parts. The first part calculates the heat transferred from the barrel to the extrudate. This part of the calculation method is based on several of Zuilichem's works [5, 20, 21, 30]. The second part of the model calculates the heat generated by viscous dissipation (Figure 2.4).

The following assumptions were made for the description:

1) Input parameters of the model are:
   a) the temperature profile imposed at the barrel,
   b) torque,
   c) feedflow.
2) Torque is converted into: pressure energy, phase transition energy and temperature increase. Hence:
   a) the torque and the heat convection are linearly related to the degree of fill,
   b) the viscosity changes of the material are described by a non-Newtonian, temperature-time dependent power law model,
   c) the amount of energy required to gelatinize the starchy food product is negligible compared to the energy consumed to increase the temperature of the product,
   d) the meltflow is highly viscous and in the laminar flow regime,
   e) only the net flow in the axial direction is considered.

A computer model, following a stepwise procedure, was developed with the following process and material variables for the input:

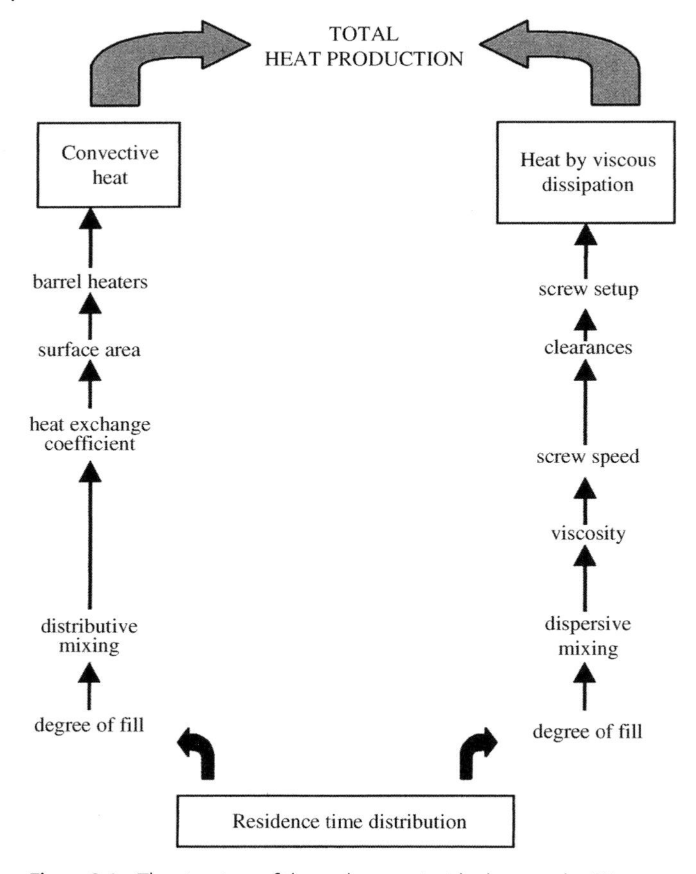

Figure 2.4 The structure of thermal energy inside the extruder [5].

- material variables: C_p, initial viscosity, melting temperature, density and moisture content,
- system variables: throughput, initial material temperature, length of extruder, screw geometry and temperature profile.

The output of this computer program is a plot of the temperature of the food product versus the axial distance in the extrusion-cooker. It is also possible to plot the temperature increase, due to penetration of heat and dissipation, versus the axial distance in the extruder.

The function of many operations in polysaccharide extrusion-cooking is to rupture the starch particles and to gelatinize their contents to a certain extent for different applications in human food and animal feed. During this process considerable changes in the rheology of the material are occurring due to the formation and breakage of gel bonds. These changes can have an impact on processing parameters and stability criteria. Janssen and van Zuilichem [31] suggested an equation especially for applications in food extrusion technology. This method is used in van Zuilichem's model and describes the effect of the mechanism of gel-formation and gel-breakage on the viscosity as a function of shear rates and fat and moisture content. They combined a power law behavior with a temperature dependence and an extra term for the formation of the network depending on the activation energy and the residence time. Now the apparent viscosity, μ_p, can be written as:

$$\mu_p = \mu_o \dot{\gamma}_p^{-n_1} e^{-a_1(MC-MC_o)} e^{-a_2 FC} e^{-b_1 \Delta T} e^{\left(\int_0^{\tau} e^{-\frac{\Delta E}{RT(t)}} Dt\right)} \quad (2.47)$$

where Dt is the convective derivative accounting for the fact that the coordinate system is attached to a material element as it moves through the extruder. With this model it became possible to explain certain instabilities as they occur during the extrusion-cooking of starch.

This model was practically verified by van Zuilichem with a co-rotating twin-screw food extruder, type MPF-50 APV Baker. A big advantage of this extruder is the modular screw design allowing one to compose almost every desired screw configuration. As a model food material, standard biscuit flour was used.

The measurements showed the accuracy of the proposed model. There is a slight deviation in calculated and measured temperature values (up to a maximum of 10 °C), giving a measured temperature slightly lower than predicted.

An advantage of the model is that it is capable of calculating the resulting temperature profile of the product over the viscosity-history as the key rheological problem. It must be said that the readings from the thermocouples at high screw speeds and relatively low throughputs are not ideal because a low degree of fill in nearly all parts of the extruder can cause the thermocouples to measure a mixture of melt temperature and steam temperature in the metering section, instead of a homogeneous product temperature.

All in all, there is a deficiency of various rheological data on biopolymers, and there is a need for further thorough examination. This will allow the development of new models which, if verified through experimental studies, will be more reliable and closer to reality.

References

1 Bruin, S., van Zuilichem, D.J., and Stolp, W. (1978) A review of fundamental and engineering aspects of extrusion of biopolymers in a single-screw extruder. *J. Food Process Eng.*, **2**.

2 Griffith, R.M. (1962) Fully developed flow in screw extruder. *Ind. Eng. Chem. Fundam.*, **1** (3), 180.

3 Harper, J.M. (1984) *Extrusion of Foods*, CRT Press Inc., Florida.

4 Janssen, L.P.B.M. (1978) *Twin-Screw Extrusion*, Elsevier, Amsterdam-New York.

5 Van Zuilichem (1992) Extrusion Cooking. Craft or Science? Ph.D. thesis, Wageningen University, Netherlands.

6 Pinto, G. and Tadmor, Z. (1970) Mixing and residence time distribution in melt screw extruder. *Polym. Eng. Sci.*, **10**, 279.

7 Dekker, J. (1976) Verbesserte schneckenkonstruction für das

extrudieren von polypropylen. *Kunststoffe.*, **66**.

8 Mościcki, L., Mitrus, M., and Wojtowicz, A. (2007) *Technika ekstruzji w przetwórstwie rolno-spożywczym (in Polish)*, PWRiL, Warsaw.

9 van Zuilichem, D.J., Swart de, I.G., and Zuiman, G. (1973) Residence time distribution in an extruder. *Lebensm. -Wiss. U. Technol.*, **6**, 184.

10 Wolf, D. and White, D.H. (1976) Experimental study of the residence time distribution in plasticating screw extruders. *AIChEJ*, **22**, 122.

11 Mościcki, L. and Mitrus, M. (2001) Energochøonność procesu ekstruzji (in Polish) Teka Komisji Motoryzacji i Energetyki Rolnictwa, PAN, vol. 1 186–193.

12 Harmann, D.V. and Harper, J.M. (1974) Modelling a forming foods extruder. *J. Food Sci.*, **39**, 1099.

13 Tsao, T.F., Harper, J.M., and Repholz, K.M. (1978) The effects of screw geometry on extruder operational characteristics. *AIChE Symp. Ser.*, **74** (172), 142.

14 Harper, J.M. (1976) Goals and activities of the Colorado State University LEC program. Low- Cost Extrusion Cookers, LEC-1. Colorado State University, Fort Collins.

15 Sahagun, J. and Harper, J.M. (1980) Parameters affecting the performance of a low-cost extrusion cooker. *J. Food Proc. Eng.*, **3** (4), 199.

16 Griffith, R.M. (1962) Fully developed flow in screw extruders, theoretical and experimental. *Ind. Eng. Chem. Fundam.*, **2**.

17 Pearson, J.R.A. (1966) Mechanical principles of polymer melt processing. *Trans. Soc. Rheol.*, **13**.

18 Yankov, V.I. (1978) Hydrodynamics and heat transfer of apparatus for continuous high speed production of polymer solutions. *Heat Transfer Soviet Res.*, **10**.

19 Yacu, W.A. (1983) Modelling of a two-screw co-rotating extruder, in *Thermal Processing and Quality of Foods*, Elsevier Applied Science Publishers, London.

20 van Zuilichem, D.J., Alblas, B., Reinders, P.M., and Stolp, W. (1983) A comparative study of the operational characteristics of single and twin screw extruders, in *Thermal Processing and Quality of Foods*, Elsevier Applied Science Publisher, London.

21 van Zuilichem, D.J., Tempel, W.J., Stolp, W., and van't Riet, K. (1985) Production of high-bolied sugar confectionery by extrusion-cooking of sucrose: liquil glucose mixtures. *J. Food Eng.*, **4**.

22 Maddock, B.H. (1959) A visual analysis of flow and mixing in extruder screws. *SPE J.*, **15**.

23 Tadmor, Z. and Klein, I. (1973) *Engineering Principles of Pacification Extrusion*, Van Nostrand Reinhold, New York.

24 Shapiro, J. (1973) Melting in plasticating extruders. PhD Thesis. Cambridge University.

25 Varmeulen, J., Scargo, P.G., and Beek, W. (1971) The melting of a crystalline polimer in a screw extruder. *Chem. Eng. Sci.*, **46**.

26 Lindt, T.J. (1976) A dynamic melting model for a single screw extruder. *Polym. Eng. Sci.*, **16**.

27 Jephson, C.H. (1953) *Ind. Eng. Chem.*, **45**, 992.

28 Ganzeveld, K.J., Capel, J.E., van der Wal, D.J., and Janssen, L.P.B.M. (1994) *Chem. Eng. Sci.*, **49**, 1639–1649.

29 Martelli, F.B. (1983) *Twin-Screw Extruders*, Van Nostrand-Reinhold, New York.

30 van Zuilichem, D.J., Jager, T., Spaans, E.J., and De Ruigh, P. (1990) The influence of a barrelvalve on the degree of fill in a co-rotating twin screw extruder. *J. Food Eng.*, **10**.

31 van Zuilichem, D.J. and Janssen, L.P.B.M. (1980) *Reology*, vol. **3**, Plenum Press, New York and London.

32 Carley, J.F. and Strub, R.A. (1978) Basic concepts of extrusion. *Ind. Eng. Chem.*, **45**, 970. 1953.

33 Metzner, A.B. and Reed, J.C. (1955) Flow on non-Newtonian fluids – correlation of the laminar, transition and turbulent flow region. *AIChEJ*, **1**, 434.

3
Raw Materials in the Production of Extrudates

Leszek Mościcki and Agnieszka Wójtowicz

3.1
Introduction

After having been forced through the forming die under pressure, extrudates expand sharply. Formation of puffed, low-density cellular materials from a hot gelatinized mass is the result of physical and chemical transformation of starchy and protein raw materials (biopolymers) into a melted mass, which, due to rapid evaporative cooling becomes a honeycomb-shaped structural composition. The most popular starchy materials used in extrusion processes are: cereal products derived from wheat, maize and rice, potato starch and dried potato grits or flakes. Other cereal products obtained from rye, barley, oats and buckwheat are used to a lesser amount, mostly in order to give nutritional enrichment or to improve the taste or functional characteristics of extrudates. Production of so-called textured proteins, referred to in the literature as TVP (texturized vegetable protein) requires high-protein raw materials responsible for formation of a fibrous structure, like grains and defatted soy meal and sunflower meal, pea or faba-bean or lupin seeds flour as well as cereals' protein fractions such as wheat gluten [1].

Table 3.1 shows a set of typical raw materials used in the recipes of popular extruded products, each of which offers a wide variety of functions:

- structure-forming,
- facilitating physical transformation during the extrusion-cooking,
- affecting the viscosity of the material and its plasticization,
- facilitating homogeneity of the dough ingredients,
- accelerating starch melting and gelatinization,
- improving the taste and color of products.

The raw materials 1–5 represent the largest and most basic group of structure-forming ingredients used in the production of snacks, crisps and TVP, the raw materials 6 and 7 create a disperse phase in structural transformation by performing the role of fillers. The raw materials 8–10 are plasticizers whose role is to reduce shearing forces and facilitate the mass flow in the extruder's barrel. The role of other

Extrusion-Cooking Techniques: Applications, Theory and Sustainability. Edited by Leszek Moscicki

ISBN: 978-3-527-32888-8

Table 3.1 The recipes of popular extruded snack products (maximum values shown in wt%).

No.	Ingredients	Maize snacks	Potato and cereal snacks	Wheat snacks
1.	Maize grits	90.0	55.0	–
2.	Wheat flour	–	–	70.0
3.	Dried potatoes (granules)	–	15.0	–
	Potato starch	–	5.0	–
4.	Soya flour (defatted)	–	–	5.0
5.	Wheat bran	–	–	10.0
6.	Wheat gluten	–	2.0	–
7.	Water	14.0	14.0	16.0
8.	Vegetable oil	1.0	1.5	1.0
9.	Emulsifier	0.3	0.3	0.3
10.	Sugar	–	–	5.0
11.	Maltodextrin	–	0.5	–
12.	Salt	1.0	1.0	1.5
13.	Flavor additives	+	+	+
14.	Baking powder	–	1.5	–
15.	Calcium phosphate	–	–	1.5
16.	Milk powder(defatted)	1.0	2.0	2.5
17.	Pigments	+	+	+

raw materials is improving flavor, color, expansion ratio and texture (e.g., crispiness or durability depending on the number and size of air bubbles in the product).

3.2 Structure-Forming Raw Materials and Additional Components

Let us take a closer look at the raw materials shown above. The most important of these are structure-forming ones, that is, those rich in starch.

Maize (US – corn) – the most popular material in the production of RTE extrusion-cooked products; appears in many varieties depending on the geographic area. Those varieties used by the food processing industry are divided into so-called hard (e.g., Argentinean *La Plata* recommended in the production of conventional corn flakes) and soft – commonly used in the production as flour and grits. Starch granules in both types of maize measure from 5 to 20 μm but display different shapes depending on the content of amylose and amylopectin. The varieties of waxy maize are characterized by low (1%) amylose content as compared with popular varieties, where its content varies from 25 to 35%. Sweet corn enjoys great popularity with its sweet tasting grains. There is also pop maize having a white endosperm with relatively high content of water, which is helpful during expanding at high-temperature (pop corn). Generally, maize cultivars have low protein content (6–10%) and these are mainly glutenins and gliadins. What follows, the transformation of the protein fractions of maize grains during the extrusion-cooking, is very similar to the behavior of wheat gluten.

Wheat – depending on the variety, starch granules in wheat seeds measure 20 to 40 μm (type A) or 1 to 10 μm (type B). Their starch fractions are relatively similar and display a stable content of amylopectin and amylose. The protein content in wheat ranges from 8 to 15%, being primarily glutenins and gliadins – the components responsible for water absorption capacity and dough elasticity. The screw or twin screws rotating in an extruder barrel cause kneading and pressing of the dough, at the same time fragmenting it into small particles. The accompanying, and sometimes too intense, heat treatment at high temperature may cause significant chemical transformation of proteins, such as deep denaturation and formation of Maillard reaction products in too great a quantity. This, in turn, may lead to a significant loss of basic amino acids, and thus decrease the nutritional value of the products. Wheat flour is another, after maize, commonly used raw material in the production of food extrudates. It is cheap, it has many features relevant from the consumer's point of view, but in terms of technological considerations it produces worse effects than maize: less expansion, less crispness and characteristic flour-like taste.

Rice – has many varieties differing from each other by the protein structure that is dependent on the morphology. Most popular rice varieties produce hard, glassy seeds with small starch granules (2–8 μm). The amylose content ranges from 15 to 27%. Waxy varieties of rice are free from amylose having only amylopectin. During the extrusion-cooking, the relatively low protein fraction (6–8%) is subject to similar changes as maize and wheat protein.

Oats – a less popular material, although very valuable owing to its nutritional value. It has recently become more widely used, especially in the production of breakfast cereals. The oat milling process to obtain oat flour requires special technology where one of the production stages is steaming, aimed at lipase inactivation. The structure of oats grain endosperm is similar to the soft wheat. Principally, the chemical composition of oats grains resembles other cereals but a distinguishing factor is a large content of fiber and oil and an increased level of globulin. Oats has the highest content of lysine compared with other types of cereals. It is important that the presence of oil in oats, although very valuable (antioxidant properties), may be the reason for disruption of the extrusion-cooking process, especially in the so-called direct high-pressure extrusion (snacks production). In this type of extrusion, a very important factor is shearing, which is a source of thermal energy, and is indeed essential in the plasticization process of the processed mass. Having these factors in mind, it is important to carefully select recipes and design an appropriate percentage of that ingredient in the mixture. This applies in particular to the production with single-screw food extruders.

Potato products – used especially for the production of pellets (semi-finished product for the manufacture of fried snacks): dried potatoes grits as well as flakes and potato starch. These products are manufactured in potato plants using a variety of cooking and drying procedures. They differ in rheological properties and performance during the extrusion-cooking process. Dried potato has a natural cellular structure of potato, filled with gelatinized starch, while the gelatinization degree may be regulated depending on the product purpose. This form is most similar to the original form of raw material, including the characteristic taste. Potato starch, most commonly added to the snack compositions, has large starch granules (60–100 μm), 20–25% amylose content and low lipid content (0.1–0.2%). In extrusion-cooking, potato starch serves as a structure-forming factor influencing improvement in the expansion of extrudates.

One of the fundamental physical properties of raw materials, having immense technological importance in extrusion, is their disintegration degree. Typical forms used in the extrusion-cooking processing are grits, semolina and flour. Each of these groups has a strictly defined grain size required by the manufacturers of extruded products. The requirements for the disintegration degree of components are directly linked to the type of production device, that is, the type of extruder: single- or twin-screw, autogenic or isothermal, and depend on the production assortment. The so-called hard materials are usually offered in the form of grits or semolina; on the other hand, soft materials are available in the fine-ground form, most commonly as flour.

Because of their construction, single-screw food extruders have limited opportunities to transport flour fractions of raw materials. The flour causes problems with appropriate filling of the space between the screw flights, and, in the case of complex mixtures, raw materials delaminating with varying disintegration degrees. Twin-screw extruders are more universal and are not subject to this type of limitation. Their work characteristic allows the use of mixtures consisting of flour fractions which are much cheaper and help maintain the homogeneity of the products. A noteworthy fact is that smaller particles of raw materials more easily exchange heat with the extruder plasticizing unit than coarse particles, such as grits, which require a higher temperature and longer heating time, and thus more energy is needed for the dough plasticization. Therefore, the optimal choice of the disintegration degree of the raw materials is a vital factor for the effective use of an extruder.

Raw materials acting as fillers. When observing extrudates under a microscope, it can be easily seen that they consist of a uniform phase of gelatinized starch, which most often contains a protein matrix, and various inclusions, namely fillers. Fillers are most frequently protein fractions of oilseeds and cereal grains added to the main ingredients. Their role is to improve the flexibility of the dough after hydration during the stage of mass plasticization inside the extruder (the effect of shear stresses caused by the screw rotations and geometry). When added in quantities from 5 to 15% by weight, they reduce the swelling of starch during the mass forcing through the die, which influences the final shape of an extruded product. The presence of fillers can clearly reduce the size of air bubbles during the expansion of extrudates; the products have a specific internal structure resembling a foam with numerous tiny holes. A positive result is also guaranteed by addition of the maize starch fraction with a high content of amylose, characterized by a higher gelatinization temperature than basic materials such as wheat or rice flour. The obtained results are similar to those following the addition of some fractions of protein; the product texture is improved, clearly the addition influences the rheological properties.

Fiber-rich components, for example, wheat bran, are also used as fillers. Their use allows direct control of the shape, texture and expansion degree of the extrudates.

This is possible due to the presence of fibers with durable cellulose cells which undergo practically no transformation during the extrusion-cooking.

Plasticizers. The plasticizers in an extrusion-cooked mixture are water, oils, fats and emulsifiers. During the extrusion-cooking process a pressure and thermal processing of biopolymers takes place consisting in, for example, rapid concentration of material being subject to shear stresses, accompanied by the transformation of mechanical energy into heat energy. The water content in the mixture ranges from 10 to 25% of the dry mass and simplifies the process by stimulating shearing, hence influencing the amount of thermal energy needed for the process. Water added in quantities above 30% of the dry mass acts as a solvent for starch polymers, thus has acting as a dispersing agent for starch granules. Moreover, the presence of microcomponents in water, such as calcium or magnesium carbonate, is of little significance.

When it comes to oils and fats, their role is twofold. First, their presence in the mixture facilitates transport and the process of material condensation inside the extruder; secondly, it has a positive effect on the quality and nutritional value of the extrudates. High temperature contributes to their easier mixing with raw materials by their absorption or collection on the surface of particles. Their lubricating functions are easily noticed during the extrusion-cooking of mixtures with low water content when, in the event of heightened viscosity, the material surface is overheated and adheres to the cylinder walls or screw; this ultimately may lead to the extruder jamming. A small addition of vegetable oil, from 0.5 to 1%, protects against degradation of carbohydrates and furthers the process of pregelatinization of starch.

On the contrary, highly disrupted raw materials, such as cereal flours, follow a different pattern. In similar conditions, the above-mentioned oil additive (0.5–1%) causes a reduction in the amount of energy, and consequently of temperature, and accelerates the mass flow in the die opening. Less shearing reduces the degradation of starch granules, which ultimately causes an increase in the viscosity of the plasticized mass. This affects the expansion ratio of extrudates, and, when the oil content is more than 3%, it may result in their compacted form free from air spaces (e.g., oats flour extrudates). In this case, to obtain a typical structure of a honeycomb in snacks or crisps, it is necessary to raise the material humidity to 30–35%. There is a limitation on the amount of fat-rich materials that should be taken into account in the recipe for such reasons as nutrition or taste. Then it is advisable to add absorbing substances, for example, bone meal or chalk. This practice is commonplace in the production of extrusion-cooked dry dog or cat food and fish feed.

Emulsifiers. These are lipid fractions with higher melting temperature used by the manufacturers of extruded products as lubricants. Due to their characteristics and effects on the processed mass, they facilitate shear and formation of a uniform extrudate surface as well as protecting against stickiness and thus making further treatment easier. The most popular emulsifiers are soy lecithin and mono- and diglyceride esters.

Raising agents. Products such as sodium phosphate (baking powder), which are added in the quantity of 1 to 2% to a mixture, provide perfect aeration of the mass and help to obtain the typical crispy and porous structure of extrudates, which is highly valued by the consumers of cereal snacks foodstuffs.

Taste components. The last, but not least, group of raw materials listed in the recipes for extruded mixtures are taste components, such as salt and sugar. Salt added in an amount of 1–1.5% easily dissolves in the dough mass, and, to a lesser extent, affects the extrusion-cooking process. If added in larger quantities, salt may influence the material pH, which sometimes can be beneficial from the technical point of view, but will not always be accepted by consumers because of the salty taste. The addition of sugar to the mixture in an amount less than 10% of the mass does not significantly affect the extrusion process either. It can be fed in the form of powder, crystal or as syrup. Higher sugar content in the mixture has a negative effect on the extrusion-cooking process, since it reduces the material temperature, which entails more excessive energy consumption for heating and lowers the degree of extrudate expansion. At its higher concentration, starch does not easily undergo transformation and the dough is more difficult to plasticize. In certain conditions, hydrolysis of sucrose can take place together with its degradation to glucose and fructose, which, when combined with protein fractions (peptides), take part in developing the color and taste during the non-enzymatic browning reaction. Simple sugars can be added to mixtures but it must be mentioned that at high temperature an excessive number of Maillard reaction products can be formed. Appropriate selection of pressure and thermal processing parameters allows almost unrestricted shape, color, and taste of the product.

3.3 Physical and Chemical Changes in Vegetable Raw Materials During Extrusion-Cooking

Extrusion-cooking is accompanied by the process of starch gelatinization, involving the cleavage of intermolecular hydrogen bonds. It causes a significant increase in water absorption, including the breakage of starch granules. Gelatinized starch increases the dough viscosity, and high protein content in the processed material facilitates higher flexibility and dough aeration [2, 3]. After leaving the die hot material rapidly expands as a result of immediate vaporization and takes on a porous structure. In the extruded dough protein membranes cover the pores, creating cell-like spaces, and starch, owing to dehydration, loses its plasticity and fixes the porous nature of the material. Rapid cooling causes the stiffening of the mass, which is typical for carbohydrate complexes embedded in a protein matrix and totally enclosed by the membrane of hydrated protein [4, 5]. The resulting product is structurally similar to a honeycomb shaped by the clusters of molten protein fibers.

As a result of combined temperature and pressure action, the processed material is subject to significant changes; the scope of these changes and their mechanism are still the subject of scientific investigations in many research institutions.

Based on the available literature and our own research, we have set out the key aspects of the chemical transition of starch and protein and the influence of the extrusion-cooking process on the nutritional value of extruded vegetable materials.

3.3.1
Changes in Starchy Materials

Starch occurs primarily in cereal grains and potatoes. It takes the form of granules of different and characteristic shape, depending on the origin and on the variety and type of fertilization. As is commonly known, two main components of starch are amylose and amylopectin, displaying different physical and chemical properties, for example, chemical structure. Amylose as a multi-D(+)glucopyranose compound linked by α-1,4 bonds has an unbranched chain structure. Amylopectin differs from amylose in that it has a larger molecular weight and numerous branched chains. These chains have the construction of amylose and are linked to the main chain by α-1,6 bonds [6].

Regardless of the variety, native starch reveals a regular crystalline structure which can be observed under a microscope. The use of a simple optical microscope allows observation of the shape and size of starch granules, specific to the starch type (Figure 3.1a). A characteristic feature of native starch is the presence of the so-called "Maltese cross", discernible as a black intersection on a shiny starch granule under polarized light (Figure 3.1b). This phenomenon is known as starch birefringence and determines the crystalline nature of starch and the degree of order in the granules [6]. If heated in the presence of water, all kinds of native starch granules lose this feature (Figure 3.1f). It is directly related to the process of starch gelatinization in which the granules swell under the influence of heat, absorb water and lose their crystalline nature.

At a temperature specific to each type of starch, known as the gelatinization temperature, starch granules irreversibly lose their regular shape and properties (such as insolubility in cold water). These changes are connected to the increase in viscosity of the heated solution and greater starch solubility in water [6]. These characteristics are an advantage in the production of the so-called functional starches used as thickeners for soups, sauces, desserts and dishes served cold. Gelatinization temperatures of popular starch types are: maize 62–80 °C, wheat 52–85 °C, potato 58–65 °C, tapioca 52–65 °C, and vary depending on the availability of water, the size and heterogeneity of the starch granules, and on the measurement method used.

Observation of the structure of extrusion-cooked maize carried with a scanning microscope [7–9] showed that if the process temperature was in the ranged 50–65 °C, the starch did not show any significant physical and chemical changes. In the case of extrusion at a temperature above 200 °C, deformation and partial or complete gelatinization of the starch granules was observed. Gelatinized starch, cellular protein and cellulose comprise one complex which decisively influences the product's expansion ability after forcing through the die [10].

Similar changes were observed in potato starch [8]. Under a microscope extrudate obtained at a temperature of 65–75 °C had visible, undamaged starch granules, especially in the external part of the product. Upon material processing at 90 °C, the starch swelled and became gelatinized, but displayed clear air chambers. Processing at 225 °C resulted in the presence of small bubbles close to the outer sides of the samples; while larger bubbles appeared in the middle of the samples. In the case of

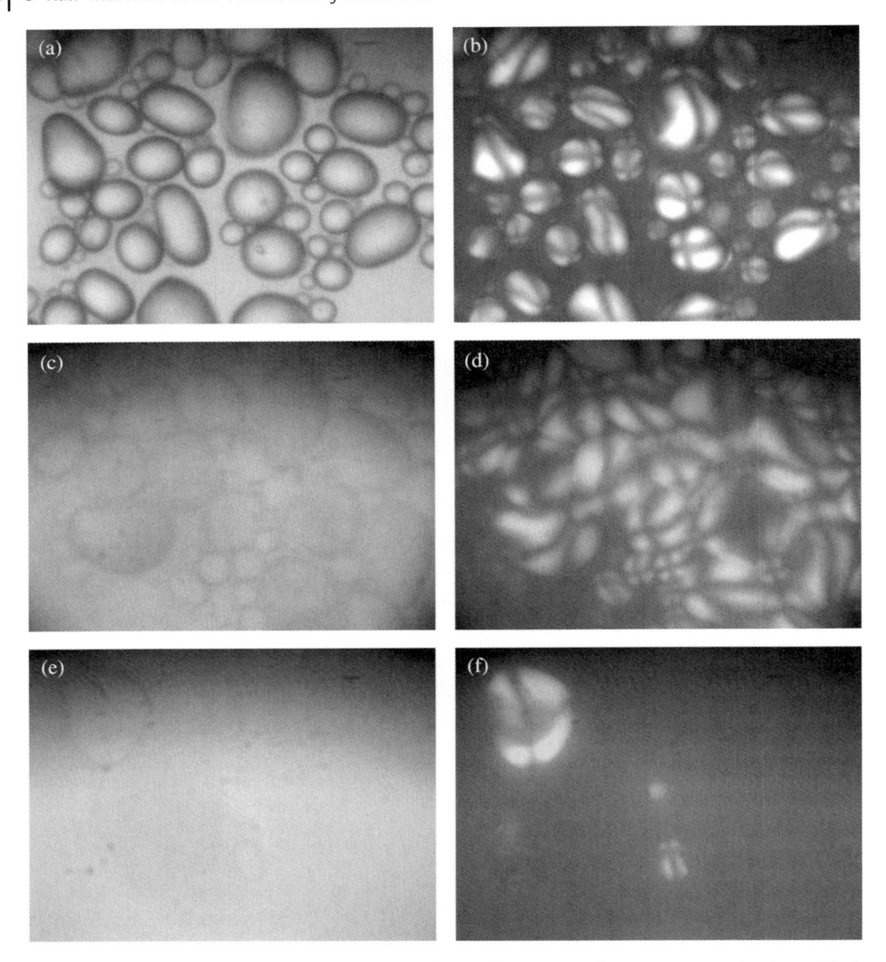

Figure 3.1 The structure of potato starch granules under an optical microscope and polarized light: (a, b) native starch in cold water, (c, d) starch heated to 50 °C, (e, f) starch heated to 90 °C.

extrusion of wheat flour [11], these phenomena were observed at temperatures of 90 to 125 °C.

It is common knowledge that unprocessed starch of cereals has the A-type structure, while the potato starch represents the structure defined as type "B". Since the structural construction of starch is altered after extrusion, and depending on the physical and chemical features of raw materials and the adopted parameters of a thermoplastic processing, the structure of extrudate clearly differs from that of the original material [12].

The influence of extrusion-cooking on the structural reorganization of starch was studied by Charbonnier *et al.* [13] using X-ray diffraction. In potato starch and manioc starch (free of lipids) as well as waxy maize starch (free of amylose) reduced crystallinity was observed at an extrusion temperature as low as 70 °C, while at higher temperatures the structure was completely destroyed, leading to an X-ray pattern typical of an amorphous state.

Very different results were obtained during the extrusion-cooking of cereal materials. For example, commercial corn starch at 22% moisture extruded at a temperature of 135 °C formed a new type of structure characterized by three main peaks in the X-ray diffraction pattern, with the major one located at 9°54′. Such a structure is similar to the butanol-amylose complex spectrum and is known as the V-amylose type of structure [9]. A similar material with lower moisture content (about 13%) and extruded at temperatures from 185 to 225 °C produced a new structure known as the extruded structure or E_V-type [12, 14]. A characteristic feature of these type of diagrams is the presence of three diffraction peaks, while the main one is shifted to 9°03′. At a temperature of 170 °C both E-type and V-type structures occurred, and by raising the initial moisture in the material to 30% processed starch exhibited the only V-type pattern.

A similar phenomenon was observed when analyzing wheat and rice extrudates processed in the presence of higher moisture content. The similarity of the two types of patterns E- and V-type, led Mercier *et al.* [8] to adopt a hypothesis that the extruded structure is helical, like the B-type, and the only differences in the two structures are the different interaxial distance in the two helixes. The difference in the interaxial distance is 1.38 nm in the V-type and 1.50 nm in the E-type. In the case of material extruded at a moisture level below 19%, the spectra were unmodified (E-type 1.5–1.47 nm, V-type 1.38–1.31 nm). At a moisture content of 30%, the previously obtained E-type spectrum was irreversibly transformed into the stable V-type. Thus, the extrusion process in cereal products produces structural changes in helical complexes with the same quantities of glucose units, rather than two separate structures.

The different behavior of potato starch, manioc and waxy maize during extrusion-cooking, under the same conditions as adopted in the thermoplastic treatment of cereals, was studied by Mercier and others [8, 12]. They assumed that during the extrusion process the structural changes in starch are stimulated by the presence of fatty acids. She added from 2 to 5% of different fatty acids to the input compositions of these raw materials. In all the experiments the same types of V-amylose patterns were obtained. The inability to form the amylose-fatty acids complexes was explained by the fact that the molecules of fatty acids used for the experiments were significantly larger than the internal space of the helical structure of amylose.

Based on other studies [15], it was found that during the extrusion-cooking process at high temperature (120–180 °C) and pressure (3–16 MPa) partial starch hydrolysis may occur. Generally, as a result of such thermoplastic treatment the milk-white starch (120–135 °C) turns dark brown (150–180 °C). At a temperature of 120–135 °C starch becomes completely gelatinized, however, no hydrolysis was observed. The combined effect of temperature and pressure causes in starch the cleavage of 1,2-glycoside bonds of sucrose, 1,4-glycoside bonds of raffinose and 1,4-glicoside bonds of maltodextrins. The bond cleavage in the molecules with higher molecular weight increases with increasing extrusion temperature and lower moisture content of the raw material [16, 17].

In extruded potato flour amylopectin chains were found to be present, while amylose chains generally disappeared. In other words, this type of thermoplastic treatment mainly destroys the α-1,4 bonds of amylose which are less numerous in amylopectin. Thus, the structural composition of amylose renders it more susceptible to breakage. In the case of cereals or potato starch enriched in fatty acids, extrusion produces a butanol-amylose complex which protects amylose fractions and prevents their destruction because of the formation of oligosaccharides in the extrudates [70].

The technological assessment of extrudates takes two factors into account: the water solubility index (WSI) and the water absorption index (WAI). These properties have been studied in many laboratories [8, 12, 13, 15, 18, 19] and the conclusions were that the WAI of many starch products increases with increasing temperature in the barrel of the extruder. It has been assumed that the maximum value is obtained in the temperature range 180–200 °C. When these temperatures are exceeded, the WAI drops causing the WSI to increase. The lower the initial moisture content in the material used in the extrusion, the higher the extrudate's WSI [20, 21]. A noticeable influence on the product properties is the percentage of amylose and amylopectin and their ratio in the processed material. For instance, operating at two extrusion temperatures (135 and 225 °C) induced WSI decreasing with increasing amylose in the raw material [18], and Matz [22], on the basis of experiments with different raw materials, stated that in order to obtain a product of proper texture and hardness it was recommended to use a raw material containing from 5 to 20% of amylopectin.

The extrusion processing of starchy materials certainly brings about changes in product viscosity (pasting characteristic) after dissolving in water. This feature is very important from the technological point of view. Research carried out with a Brabender viscometer showed that the characteristic viscosity curve for starch is clearly reduced through extrusion; at the same time, the decrease in viscosity is greater if higher temperatures are applied during the extrusion-cooking [12]. The application of higher pressure during the extrusion (compression changing) did not affect the extrudate viscosity; however, it affected the viscosity stability of products retained at a temperature of 95 °C. The final viscosity of extrudate tested at a temperature of 50 °C was significantly lower than that of the starch materials. In some cases, the properties of the extrudate may be determined by amylose bonding with fatty acids or monoglycerides [23].

Another factor determining changes in the starch molecules during the extrusion-cooking process is the pressure and the values of existing shearing forces [24]. In order to obtain certain technological properties of extrudates, which are often semi-finished products intended for further processing, it is necessary to set proper parameters for the extrusion process. This is achieved by the use of screws with varying compression degrees, the relevant rpm of the working screw, the appropriate die size, SME input, and so on.

An interesting and important extrudate feature is the digestion rate of starch. Research on starch digestibility processed by extrusion-cooking has shown [18] that it mostly depends on the structure generated. The researchers concluded that maize starch extruded at a temperature of 135 °C was significantly less digestible if coarse.

Fine ground extrudate treated with α-amylase displayed a 20% higher degree of enzymatic hydrolysis than non-ground material and, after grinding it into flour, the hydrolysis degree reached 80%. Such differences in digestion were not present in extrudates obtained after the application of process temperature of 225 °C. In addition, with such thermal treatment, the highest expansion of extruded maize was observed.

The lipid-amylose complexes formed during extrusion inhibit the activity of α-amylase enzyme. At 225 °C, amylose significantly crystallizes, as confirmed by X-ray diffractometry. The complexes developed as a consequence of chemical linking amylose with fatty acids during the extrusion may impact negatively on the extrudates' digestibility; however, these changes are not of major importance [25]. Generally, the extrusion-cooking process does not have a significant impact on the *in vitro* digestibility of starchy products. There have been experiments [19, 25, 26] showing an increase in bioavailability of starch extrudates by organisms. Extruded feed, for example, urea–maize grits mixtures [27], was successfully applied in the feeding of ruminants.

3.3.2
Changes in Protein-Rich Materials

Thermoplastic processing of vegetable materials by extrusion-cooking also causes significant changes in the protein substances which can be found in vegetable materials.

The mechanism of obtaining textured vegetable protein by extrusion has been investigated by many researchers. Taronto *et al.* [28] found that extrudate obtained from soya meal and cotton seeds contained proteins of diverse structure and dispersed carbohydrates in an amount dependent on the material used. Analyzing this process from the chemical point of view, protein texturization would consist in the stretching and straightening of polypeptide structures preceded by the cleavage of bonds which are responsible for the tertiary and quaternary protein structure [3]. As a result of such thermoplastic treatment, the hydrophilic character of vegetable protein is lowered [29, 30]. As confirmed by numerous studies, temperature is considered a priority in these processes [31, 32]. Soya-maize extrudates obtained at a temperature of 121 °C showed a nitrogen solubility index (NSI) of 42%; 25% lower than with the original material. The NSI of extrudate obtained at 149 °C was 16.6%. Lowering of the NSI was also observed with material that was more finely ground before extrusion [33].

Jeunink [34]; [35] believes that not only do temperature and the type of material play an important role in the process, but also the physical and chemical properties of particular components of the extruded mixture are important. The decreasing NSI value is probably associated with the loss of covalent sulfur bonds, and disulfide bridges of sulfuric amino acids with simultaneous decrease in the cysteine + cystine content in the protein of extruded beans.

Burgess and Stanley [36] assumed that in the processes of structural protein transition intermolecular isopeptide bonds may also be of significance. When examining extruded wheat flour and maize flour, it was observed that the extrusion process influenced the reduction in the level of albumin, globulin, prolamin and gluten in the final product. Only the extruded sorghum flour displayed higher prolamin content in relation to the input material. Linko *et al.* [37] demonstrated that, with proper setting of the parameters of thermoplastic processing, it is possible to obtain cereal extrudate of a relatively high amylolytic activity. Nevertheless, the influence of extrusion on the enzymatic activity must be taken into account, particularly when it is important from the technological point of view (the purpose of the product), or when these processes are carried out in order to explore the mechanism of starch dextrination that occurs during extrusion.

3.3.3
Changes in Fibers

The role of fiber in dietetics is widely known and appreciated, especially its auxiliary function in digestive processes and its positive impact on the vermicular movement. Polysaccharides and lignin – the basic fiber components – perform differently during baro-thermal treatment such as extrusion [38]. The degree of fiber degradation depends on the size of the shear stresses. Björck and Asp [39] found that extrusion processing almost doubled the content of water-soluble fiber in processed wheat grain. It was simultaneously tested by Varo *et al.* [40] who used many different analytical methods to test the cereal extrudates. There were no statistically significant changes in the quantity of fiber in relation to the input material before extrusion.

The positive effect of fiber in the diet of patients with diabetes is well-substantiated in the medical literature [41, 42]. Nygren *et al.* [43] led nutritional research involving bran-wheat extrudates and reaffirmed their suitability for diabetics. Such products have been recently recommended by dieticians and gastroenterologists.

An interesting subject related to the decomposition of fiber to glucose and lignin, combined with the use of urea during the extrusion of pine sawdust for the purpose of feed, was studied in the Netherlands [44]. The obtained results confirmed the usefulness and effectiveness of the extrusion technique in the process of cellulose degradation, and the final product, energy- and protein-rich, has proved to be a good alternative as a valuable feed component for ruminants which have the capacity to absorb proteins of inorganic origin.

3.3.4
Changes in Vitamins

Vitamins can be destroyed through the action of temperature or by oxidation. Since extrusion mostly involves thermal treatment at temperatures of 100 °C or higher, some loss of vitamins in the processed material is expected, especially of the temperature-sensitive and water-soluble ones, such as vitamin C.

Many authors have reported studies on these matters and confirmed the occurrence of this phenomenon, but because of the HTST-type shock treatment, they found that the extent of losses was much less compared with conventional methods such as static, long-term cooking. Since the basic raw materials used in the processing of extruded products are cereals, much attention was attached to the changes in the B-group vitamins, particularly thiamine B_1, riboflavin B_2 and niacin [45, 46]. Thiamine, being the most sensitive to temperature, is damaged during extrusion, depending on the processing conditions, temperature rise and screw speed. Riboflavin losses are much lower (retention post-extrusion was 92%) and decrease with increasing water content in the mixture [47].

Anderson *et al.* [48] carried out extrusion of cereal flakes in a twin-screw extruder and found that the initial moisture content of the raw material has a far greater impact on the thiamine depletion than the increase in processing temperature. In a mixture with a moisture content of 14% extruded at a temperature of 150 °C, thiamine losses ranged from 60 to 90%; after adding water to the raw material to reach a moisture content of 30% or more, thiamine losses decreased to only a small percentage at a processing temperature of 200 °C. Riboflavin and niacin in the above-mentioned studies showed high resistance to the conditions of the process, as seen in their limited degradation in extrudates.

The losses of vitamin C in the extruded raw materials usually do not exceed 80%; with adjustment and careful application of suitable extrusion parameters, the losses can be reduced by up to 50–60% [32, 49, 50]. Minor loss is observed during extrusion of low moisture content materials. The most advantageous solution is to enrich the extrudates (e.g., by spraying) after the extrusion process [2]. This approach is much employed by the manufacturers of extruded cereal foodstuffs who add a number of other micronutrients at this stage. Table 3.2 shows collective data on the changes in the basic raw materials components taking place during the extrusion-cooking their dependence on the adopted process conditions.

Vitamin A reveals high stability in extruder products, much higher than in the case of raw materials. According to researchers, it is associated with an increased susceptibility to the extraction of fat-soluble vitamins after baro-thermal proces-

Table 3.2 The effect of extrusion conditions on the changes of nutrients in extrudates (where: + increase, – loss, o no effect, * high temperature, ** high temperature, low moisture content) [39].

Process components	Components Protein	Starch		Vitamin preservation			
	Lysine	Gelatini-zation	Dextrination	B_1	B_2	C	A
Temperature	–	+	+	–o	+o	–	o
Moisture content	+	+*		+	–o	–**	
Screw rpm	–o	–		–o	–	–	+
Screw geometry	–		+	o		–	
Die diameter	+	–		+o	o	+	
Torque, extrusion pressure	–		+			o	

sing [46, 51]. The same is true for vitamin E which is almost not destroyed during extrusion.

3.4
Nutritional Value of Extrusion-Cooked Foods

Temperature treatment of vegetable material containing proteins and reducing sugars usually leads to a deterioration of the nutritional characteristic of proteins. This phenomenon is primarily due to Maillard reactions [52, 53]. As a result of a reaction between the free amino acids of proteins and aldehyde clusters of sugars, enzyme-resistant bonds are developed. The formed complexes can be further transformed while dark colored melanoids appear and amino acids become broken; the most considerable losses are observed in lysine, histidine, threonine, phenylalanine and tryptophan, the deficit amino acids in food [54].

Moderate heating of vegetable materials generally improves the nutritional value of their protein, yet too intense thermal treatment has an adverse effect. Hence, temperature is one of the key factors affecting the biological value of the processed materials.

The impact of extrusion on the loss of lysine in extrudates has been tested by many authors [32, 37, 55–57]. Jokinen *et al.* [58] found that the process of thermoplastic processing, similar to that taking place during extrusion-cooking, can be divided into three steps:

- initial phase (up to 1 min), characterized by rapid reduction in the level of lysine;
- temporary phase;
- final phase, in which the content of lysine is subject to minimal fluctuation.

Since the particle residence time distribution in an extruder ranges from 10 to 70 s [59, 60] the most intensive lysine transition takes place in the initial phase. In his model research, Wolf [57] saw the maximum drop in lysine level in the processed material mainly during the treatment in isothermal conditions, without any movement of material in the device. Extrusion-cooking, due to process characteristics (intensive mixing and consolidation), despite the high temperatures applied, has a small effect on the reduction of the amount of lysine in the final product. Particularly good results are obtained in the HTST type of extrusion, which relies on a very rapid thermoplastic treatment of raw materials [10, 26, 56].

The hazard of Maillard reaction products formation related to the thermal processing of proteins can be regulated in the extrusion process by suitable adjustment of the following parameters:

- temperature, which should be kept to a minimum,
- screw rotations (an increase in rpm raises the temperature in the cylinder),
- water content in the processed material.

Mercier *et al.* [8], producing children biscuits from wheat flour and sucrose with the addition of powdered milk, noted that in the extrusion of such a mass at 15%

moisture content and at the temperature of 200 °C Maillard reaction products formation is relatively easily expedited. The result was a 50% reduction in lysine and methionine content and a decline in sucrose (18%) and lactose (47%). The disintegration of amino acids and reactions of proteins with sugars were kept to a minimum when the moisture content of the raw material was increased to 36%. Similar results were obtained in other studies [10, 61], when, upon selecting adequate material moisture content before extrusion, the loss of amino acids in cereal extrudates decreased to 10%.

The processes of carbohydrate hydrolysis occurring during extrusion can explain the energy stimulating transformations of protein amino acids. Lactose is more susceptible to hydrolysis than sucrose; this is the reason for the rapid browning of the product during extrusion. The replacement of sucrose with fructose resulted in 80% loss of lysine in the cereal extrudates compared with the input mixture [62].

Many experiments have proved that the extrusion processing of vegetable materials has no major impact on *in vitro* protein digestibility, although the amount of amino acids was reduced [33, 63]. The effect of temperature and water content in cereal materials on *in vitro* digestibility of extrudates was studied by Camus and Laporte [64]. They noted that with increase in the process temperature to 225 °C there is a growing susceptibility of the products to pepsin action (at 14% water content). With higher material moisture content, pepsin proteolytic action decreased. The degree of proteolysis also depends on the type of extrudate. If a soya–maize mixture was used, the effect of temperature was much lower. Proteolysis with trypsin was dependent on the degree of inactivation of trypsin inhibitors, on the presence of Maillard reaction products and the availability of amino acids, determining the activity of the proteolytic enzymes.

The temperature of extrusion-cooking is sufficiently high to inactivate thermolabile agents inhibiting digestion and enzymes. In the case of legumes, the pea flavor is also largely neutralized [26, 65, 66]. When extruding mixtures of legumes and cereals, Sautier and Camus [67] ascertained a total inactivation of trypsin inhibitors, however, even when treated at a temperature of 200 °C, they noticed the presence of galactosides and glycosides in products.

Research on the protein efficiency ratio (PER) of vegetable extrudates showed that the PER value is generally comparable to that for casein (PER for casein – 2.5) and, depending on the material, is in the range 2.2 to 2.5 [50, 55, 67, 68]. The extrusion of soya–maize or soya–sorghum mixtures or each of the ingredients separately, does not affect the PER value of the finished product [68]. On the other hand, Muelenaere and Buzzard [69] demonstrated that the PER was positively influenced by the degree of material disintegration, which was associated with the residence time distribution of particles inside the barrel.

Sautier and Camus [67] performed studies on the *in vitro* and *in vivo* digestibility of extruded bean–maize mixture and did not observe any gastric disorders with animals, or metabolism differences or food assimilation in relation to a control group fed with conventional feed. Similar results were obtained by the same authors among volunteers who consumed extrudates added to food in various forms.

3.5 Concluding Remarks

For several years a mounting volume of vegetable products used for the production of food and feed and obtained through the extrusion-cooking technique has been noted. A growing number of published patents and scientific papers prove that extrusion has permanently settled into agricultural and food processing.

Starch extrudates, due to their high stability of viscosity, are widely used in the manufacture of starch modifiers and functional substances. In bakery, extruded maize is used as a binding agent for meat fillings and as a raising agent in baking cakes to improve their texture and taste. The physiochemical properties of starch extrudates (low solubility and viscosity) make such products very competitive if compared with those obtained through chemical modifications. Extrusion of lipid-free starch increases its digestibility as a result of the formation of oligosaccharides – this determines the usefulness of this technique in baby food manufacture [5].

The greatest interest in the extrusion-cooking technique is in the production of cereal snacks, breakfast cereals and textured proteins obtained from soya beans and used as animal protein replacements in food processing. Although soy proteins play a key role, experiments have been performed with other vegetable materials such as broad bean, faba bean or chickpea, which can be substitutes for this plant.

It is also worth noting that extruded products demonstrated a long shelf life. Partial lipid bonding by the processed mass increases their resistance to oxidation processes, while lipases and lipooxidase are almost completely inactivated. In addition, thermal processes cause deep sterilization, therefore microbiological contamination of the obtained products is limited. If proper storage conditions are available, these products can be stored for more than a year.

The utilitarian importance of the extrusion-cooking of vegetable raw materials has made it widely implemented for food and feed manufacture in Poland. The manufacture of adapted twin-screw extruders has brought significant economic benefits. These machines are more efficient and less energy consuming than single-screw extruders. Extrusion shows great potential for further expansion in the domestic food and feed industry. This is not only on account of its growing application abroad, especially in developed countries. Extrusion is a giant step in the use and marketing of those products which are regarded as having little or no economic value, such as the seeds of leguminous plants.

References

1 Mercier, C. and Cantarelli, C. (1986) Pasta and extrusion cooked foods. Some technological and nutritional aspects, in *Proceedings of an International Symposium in Milan, Italy*, Elsevier Applied Science Publishers Ltd, London.

2 Mościcki, L., Mitrus, M., and Wójtowicz, A. (2007) *Technika ekstruzji w przemyśle rolno-spożywczym (in Polish)*, PWRiL, Warszawa.

3 Rutkowski, A. and Kozłowska, H. (1981) *Preparaty żywnościowe z białka roślinnego (in Polish)*, WNT, Warszawa.

4 Cheftel, J.C. (1986) Nutritional effects of extrusion cooking. *Food Chem.*, **20**, 263–283.

5 Hauck, B.W. (1979) Future marketing opportunities for extrusion cooked cereals. 64th Annual Meeting American Association of Cereal Chemistry, Washington DC.

6 Thomas, D. and Atwell, W. (1999) *Starches*, American Association of Cereal Chemistry, St. Pauls, Minnesota, USA.

7 Mercier, C. and Feillet, P. (1975) Modification of carbohydrate components by extrusion cooking of cereal products. *Cereal Chem.*, **52**, 283–297.

8 Mercier, C., Charbonniere, R., Grebaut, J., and de la Gueriviere, J.F. (1979) Structural modyfication of various starches by extrusion-cooking with a twin-screw french extruder, in *Polysaccharides in Food* (eds J.M.F. Blanshard and J.R. Mitchell), Butterworths, UK.

9 Mercier, C., Charbonniere, R., Grebaut, J., and de la Gueriviere, J.F. (1980) Formation of amylose-lipid complexes by twin-screw extrusion cooking of manioc starch. *Cereal Chem.*, **57** (1), 4–9.

10 Chiang, B.Y. and Johnson, J.A. (1977) Gelatization of starch in extruded products. *Cereal Chem.*, **54**, 436–443.

11 Kim, J.C. and Rottier, W. (1980) Modification of aestivum wheat semolina by extrusion. *Cereal Food Worlds*, **24**, 62–66.

12 Mercier, C., Linko, P., and Harper, J.M. (1989) *Extrusion Cooking*, AACCH, Inc., St. Paul, Minnesota, USA.

13 Charbonniere, R., Duprat, F., and Guilbot, A. (1973) Changes in various starches by cooking extrusion processing. *Cereal Sci. Today*, **18**, 286.

14 Cumming, D.B., Stanley, D.W., and De Man, J.M. (1972) Fate of water soluble soya protein during thermoplastic extrusion. *Can. Inst. Food Sci. Technol. J.*, **5**, 124.

15 Lorenz, K. and Johnsson, J.A. (1972) Starch hydrolysis under high temperatures and pressures. *Cereal Chem.*, **49**, 616–628.

16 Mościcki, L. (2002) Zmiany właściwości fizykochemicznych surowców roślinnych poddawanych procesowi ekstruzji. *Przegląd Zbożowo-Młynarski*, **6**, 27–29 (in Polish).

17 O'Connor, C. (1987) *Extrusion Technology for the Food Industry*, Elsevier Applied Science Publisher Ltd, USA.

18 Anderson, R.A., Conway, H.F., and Pepliński, A.J. (1970) Gelatinization of corn grits by roll cooking, extrusion cooking and steaming. *Die Starke*, **22**, 130–135.

19 Mościcki, L. (1980) Badania wybranych cech jakościowych ekstruderatów pochodzenia roślinnego Biuletyn Informacyjny Przemysłu Paszowego, 4, 37 (in Polish).

20 Harper, J.M. (1981) *Extrusion of Foods*, vol. 1 & 2, CRC Press Inc., Florida, USA.

21 Quing, B., Ainworth, P., Tucker, G., and Marson, H. (2005) The effect of extrusion conditions on the physiochemical properties and sensory characteristics of rice-based expanded snacks. *J. Food Eng.*, **66**, 283–289.

22 Matz, S.A. (1976) *Snack Food Technology*, AVI publishing, Wesport, Conn.

23 Mercier, C., Charboniere, R., Gallant, D., and Guilbot, A. (1980) Structure and digestibility alterations of cereal starches by twin-screw extrusion cooking, in *Food Process Engineering*, vol. 1, *Food Processing Systems* (eds P. Linko, Y. Mälkki, J. Olkku, and J. Larinkari), Applied Science Publishers Ltd, London, pp. 795–807.

24 Meuser, F., Pfaller, W., and Van Lengerich, B. (1987) Technological aspects regarding specific changes to the characteristic properties of extrudates by HTST-extrusion cooking, in *Extrusion Technology for the Food Industry*, Elsevier Applied Science Publishers Ltd, London, pp. 35–54.

25 Vermorel, M. (1974) Influence du traitement d'extrusion sur la digestibilite de la semuele de mais et de la fecule de pomme de terre sur le rat en croissance Rapport INRA.

26 Smith, O.B. (1975) Engineering 'meat'. *Food Eng.*, **47**, 48.

27 Mościcki, L. and Zuilichem, D.J. (1984) Animal feed applications of extrusion-cooking and a Polish example, in *Extrusion Cooking Technology*, Elsevier Applied Sci. Publishers, London, pp. 129–141.

28 Taronto, M.V., Meincke, W.W., Cater, C.M., and Mattil, K.F. (1975) Parameters affecting the production and character of extrusion texturized defatted glandless cottonseed meal. *J. Food Sci.*, **40**, 1264–1269.
29 Harper, J.M. (1986) Extrusion texturization of foods. *Food Technol.*, **40**, 70–76.
30 Mościcki, L. (1980) Badania nad procesem teksturyzacji soi metodą ekstruzji. *Roczniki Nauk Rolniczych*, **t.74-C-4**, s.54 (in Polish).
31 Aguilera, J.M. and Kosikowski, F.V. (1978) Extrusion and roll-cooking of corn-whey mixtures. *J. Food Sci.*, **43**, 225–230.
32 Lorenz, K., Jansen, G.R., and Harper, J. (1980) Nutrient stability of full-fat soya flour and corn-soy blends produced by low-cost extrusion. *Cereal Foods World*, **25**, 161–172.
33 Maga, J.A. (1978) Cis-trans fatty acid ratios as influenced by product and temperature of extrusion cooking. *Lebensm. -Wiss. U. – Technol.*, **11** (4), 183–184.
34 Jeunink, J. (1979) Modification chemiques et physico-chemiques de proteines vegetables texturees par cuisson-extrusion. Doct. Ing. De l'Universite des Sciences et techniques du Languedoc, Montpellier.
35 Jeunink, J. and Cheftel, J.C. (1979) Chemical and physicochemical changes in field bean and soya bean proteins texturized by extrusion. *J. Food Sci.*, **44** (5), 1322–1325.
36 Burgess, L.D. and Stanley, D.W. (1976) Research note – a possible mechanism for thermal texturization of soya bean protein. *Can. Inst. Food Sci. J.*, 228–231.
37 Linko, P., Colonna, P., and Mercier, C. (1982) High-temperature, short-time extrusion cooking, in *Advances in Cereal Science and Technology*, vol. IV (ed. Y. Pomeranz), American Association of Cereal Chemists Inc., St. Paul, pp. 125–235.
38 Huth, M., Dongowski, G., Gebhardt, E., and Flamme, W. (2000) Functional properties of dietary fibre enriched extrudates from barley. *J. Cereal Sci.*, **32**, 115–128.
39 Björck, I. and Asp, N.-G. (1984) Effects of extrusion cooking on the nutritional value, in *Extrusion-Cooking Technology*, Elsevier, London.
40 Varo, P., Verlaine, K., and Koivistoinen, P. (1984) The effect of heat treatment on dietary fibre contents of potato and tomato. *J. Food Science & Technology*, **19**, 485–492.
41 Brodribb, A.J. and Humphreys, D.M. (1976) Diverticular disease: three studies, Part III: Metabolic effects of bran. *Br. Med. J.*, **1**, 428–430.
42 Jenkins, D.J.A. (1979) Dietary fibre, diabetes, and hyperlipidaemia. Progress and prospects. *The Lancet*, **15**, 1288–1290.
43 Nygren, C., Hallmans, G., Jonsson, L., and Asp, N.-G. (1982) Effects of processed rye and wheat bran on glucose metabolism. International Symposium on Fibre in Human and Animal Nutrition, Palmerston North, New Zealand, abstract no. 176.
44 Mościcki, L. and Zuilichem, D.J. (1986) Ekstrudowane trociny - pasza dla przeżuwaczy. Biuletyn Informacyjny Przemysłu Paszowego, 51–56 (in Polish).
45 Beetner, G., Tsao, T., Frey, A., and Lorenz, K. (1976) Stability of thiamine and riboflavin during extrusion processing of tricticale. *J. Milk Food Technol.*, **39** (4), 244–245.
46 Zielinski, H., Kozłowska, H., and Lewczuk, B. (2001) Bioactive compounds in the cereal grains before and after hydrothermal processing. *Innovative Food Science & Emerging Technologies*, **2**, 159–169.
47 Camire, M.E., Camire, A., and Krumhar, K. (1990) Chemical and nutritional changes in foods during extrusion. *Food Sci. Nutrition*, **29**, 35–57.
48 Andersson, Y., Hedlung, B., Jonsson, L., and Svensson, S. (1981) Extrusion cooking of a high-fiber cereal product with crispbread character. *Cereal Chem.*, **58** (5), 370–374.
49 Maga, J.A. and Sizer, C.E. (1978) Ascorbic acid and thiamin retention during extrusion of potato flakes. *Lebensm. Wiss. U. -Technol.*, **11** (4), 192–194.
50 Muelenaere, H.J.H. and Buzzard, J.C. (1969) Cooker extruders in service of world feeding. *Food Technol.*, **23**, 71–77.

51 Lee, T., Chen, T., Alid, G., and Chichester, C.O. (1978) Stability of vitamin A and provitamin A in extrusion cooking processing. *AICHE Symp. Ser.*, **74** (172), 192–195.

52 Cheftel, J.C. and Bjorck, I. (1979) Maillard reaction during extrusion-cooking of protein enriched biscuits. International Symposium on Maillard Reactions In Food, Uddevala, Sweden.

53 Lempka, A. and Kasperek, M. (1977) *Związki Chemiczne Produktów Spożywczych*, PWN, Warsaw (in Polish).

54 Pongor, S. and Matrai, T. (1976) Determination of available methionine and lysine in heat treated soya bean samples. *Acta Alimentaria*, **5** (1), 49–55.

55 Molina, M.R., Bressani, R., Cuevas, R., Gudiel, H., and Chauvin, V. (1978) Effects of processing variables on some physiochemical characteristics and nutritive quality of high protein foods. *AICHE Symp.*, **74**, 172. 153–157.

56 Tsao, T.F., Frey, A.L., and Harper, J.M. (1978) Available lysine in heated fortified rice meal. *J. Food Sci.*, **43**, 1106–1108.

57 Wolf, J.C. (1978) Comparisons between model predictions and measured values for available lysine losses in a model food system. *J. Food Sci.*, **43**, 1486–1490.

58 Jokinen, J.E., Reineccius, G.A., and Thompson, D.R. (1976) Losses in available lysine during thermal processing of soya protein model system. *J. Food Sci.*, **41**, 816–819.

59 Olkku, J. (1980) Residence time distribution in a twin-screw extruder, in *Food Process Engineering*, vol. 1, 791–794.

60 Zuilichem, D.J., Swart, J.G., and Buisman, G. (1973) Residence time distribution in an extruder. *Lebensn. -Wiss. U. -Technol.*, **6**, 184–188.

61 Beaufrand, M.J., de la Guériviere, J.F., Monnier, C., and Poullain, B. (1978) Influence du procédé de cuisson extrusion sur la disponibilité des proteines. *Ann. Nut. Aliment.*, **32**, 353–364.

62 Li Sui Fong, J.C. (1978) These Docteur de Specialite en Sciences Alimetaires, Montpellier.

63 Conway, H.F., Lancaster, E.B., and Bookwalter, G.N. (1968) How extrusion cooking varies product properties. *Food Eng.*, **41**, 102–104.

64 Camus, M.C. and Laporte, J.C. (1980) Proteolyse in vitro de casein et de gluten par les enzymes pancreatiques. *Reprod. Nutr. Develop.*, **20** (4A), 1025–1039.

65 Gonzalez, Z. and Perez, E. (2002) Evaluation of lentil starches modified by microwave irradiation and extrusion cooking. *Food Res. Int.*, **35**, 415–420.

66 Mościcki, L. (1982) Zmiany w surowcach roślinnych poddawanych procesowi ekstruzji. *Postępy Nauk Roln.*, **5**, 90–104 (in Polish).

67 Sautier, C. and Camus, M.C. (1976) Valeur nutritionelle et acceptabilité chez l'homme de proteines végétales texturées. *Rev. Franc. Corps. Gras*, **23**, 203–208.

68 Jansen, G.R., Harper, J.M., and O'Denn, L. (1978) Nutritiona evaluation of blended foods made with a low-cost extruder cooker. *J. Food Sci.*, **43**, 912–915.

69 Mustakas, G.C., Albrecht, W.J., Bookwalter, G.N., McGhee, J.E., Kwolek, W.F., and Griffin, E.L. (1970) Extruder processing to improve nutritional quality, flavor and keeping quality of full-fat soya flour. *Food Technol.*, **24**, 1290–1296.

70 Mercier, C. (1977) Effect of extrusion cooking on potato starch using a twin-screw French extruder. *Die Stärke*, **28** (2), 48–52.

4
Production of Breakfast Cereals

Leszek Mościcki and Andreas Moster

4.1
Introduction

Cereal flakes (including corn flakes), muesli type products, small direct extrusion-cooked snacks in the form of balls, rings or shells are currently the most popular products known as breakfast cereal foodstuffs or more concisely as breakfast cereals (Figure 4.1). These products have gained enormous popularity in developed countries owing to their taste and dietetic features. They are easy to use and do not require cooking, as was immediately recognized by the consumers, especially children [4, 8].

In 2009 the consumption of this type of product in EU countries was estimated at just circa 3 kg per capita. However, in Central Europe it achieved the over 1 kg per inhabitant, which is still a relatively small volume; yet, the growth of this market segment has been very impressive in recent years (Figures 4.2–4.4).

Corn flakes have gained the greatest popularity worldwide and today are the most sought-after breakfast food in this group of products. They were also the first products of this sort in Western Europe, which in the 1970s broke the trail for the finished cereal products professionally referred to as RTEs (ready-to-eat). A dynamic development in this area of food processing in Central and Eastern Europe started in the 1990s after political transformation. Today's local markets' offer is broad and varied and, first and foremost, manufactured by domestic producers. Nowadays, local producers and importers offer an increasingly wider range of breakfast cereals in attractive shapes, coated with various icings, enriched with additives such as: filling (co-extruded pillows), micronutrition, vitamins and dried fruit. The demand for such food is surging and is estimated at 10–15% annually in the coming years. Taking into account Poland as a good example, despite the enormous qualitative and quantitative progress (in recent years few modern production lines have been launched), there is still much to do in the local market, which falls behind most EU countries [11].

Corn flakes are produced by two methods: traditional and extrusion-cooking (indirect method) [1–5, 9]. The traditional method consists of many hours' cooking of cut grains (like maize grits for the real traditional corn flakes) and, despite high

Extrusion-Cooking Techniques: Applications, Theory and Sustainability. Edited by Leszek Moscicki

ISBN: 978-3-527-32888-8

Figure 4.1 Popular assortment of breakfast cereals.

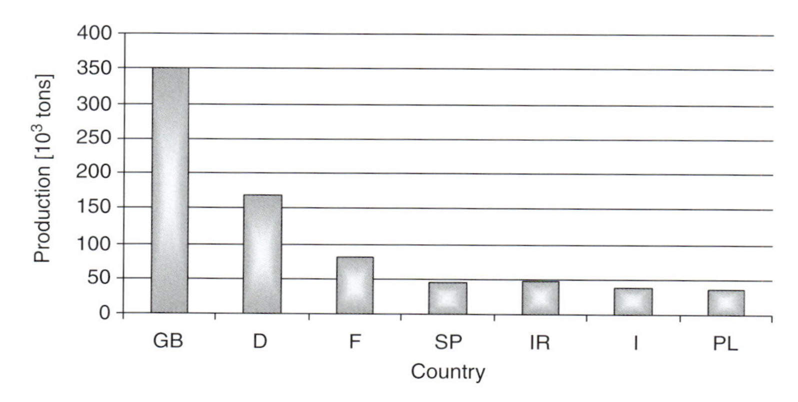

Figure 4.2 Production of breakfast cereals in European countries in 2005 [11].

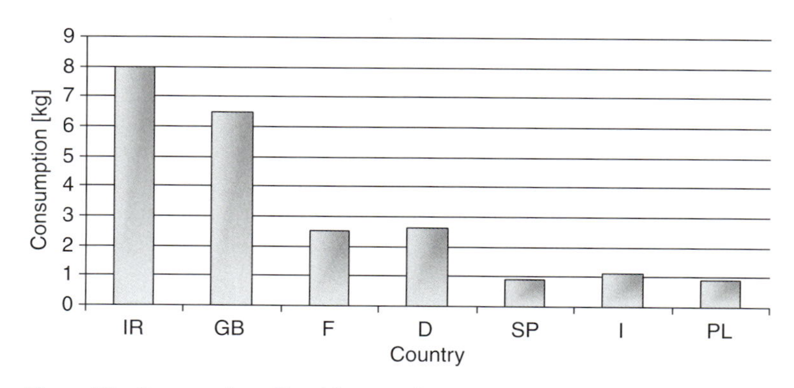

Figure 4.3 Consumption of breakfast cereals per capita in European countries in 2005 [11].

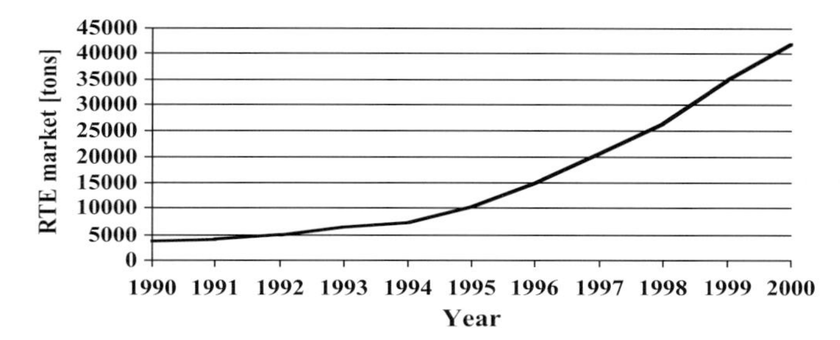

Figure 4.4 The growth dynamics of the breakfast cereal market in Poland in the years 1990–2000 [11].

energy consumption and labor load (see Figure 4.5), was still the leading technology of production in many countries of the world. In the past it was determined by several factors the most important of which are: tradition and expensive equipment at the manufacturers' disposal and the ability to preserve the distinctive flavor, color and

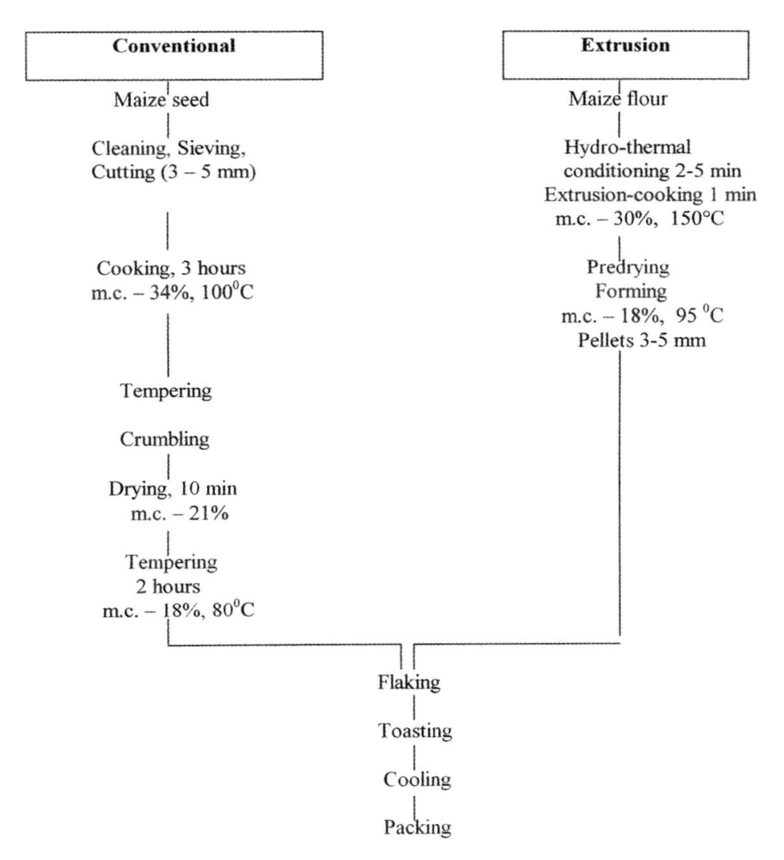

Figure 4.5 A schematic review of the production processes in the conventional and modern manufacture of corn flakes.

texture of flakes that consumers are accustomed to. The emergence of the extrusion technique revolutionized, in a sense, the production of flakes which today can be produced much more cheaply, are more regular in shape but also brighter and slightly different in quality; today's extruded flakes come nearer and nearer to the traditional ones. They lack the flavor and bite, which connoisseurs recognize with ease, and absorb more moisture, sometimes very quickly (fluff in milk). This happens when the starch is gelatinized too much or even destroyed (too much shear by extrusion) [1–9].

The key benefits of the implementation of the extrusion technique are not only lower labor demand and cost-effectiveness of production but above all the opportunity to manufacture a wide range of cereal flakes of RTE type (not only maize) obtained from flour and not from grain. Furthermore, using extrusion a product changeover can be done relatively quickly and with much less effort. This has led to the broadening of the assortment identified as breakfast foodstuffs in recent years, including multi-ingredient products of attractive shape, taste and color.

4.2 Directly Extruded Breakfast Cereals

The production of directly extruded breakfast cereals in the form of small snacks shaped like cereal seeds, balls, rings or shells and special new shapes (Figures 4.6–4.8.) is performed on a typical set of machinery consisting of a weighing system (to have the right recipe), a mixer, a high pressure extruder with an adequate die-head and a drier with cooler. Optionally a coating by spray coating drum is possible for a better look and taste or for the addition of vitamins [1, 6].

4.3 Flaked Cereals

An early improvement in the conventional "cooking" system, accomplished through extrusion, was the replacement of much of the post-cooking processing with simple

Figure 4.6 Different samples of direct expanded breakfast cereals (uncoated balls and rice crispies).

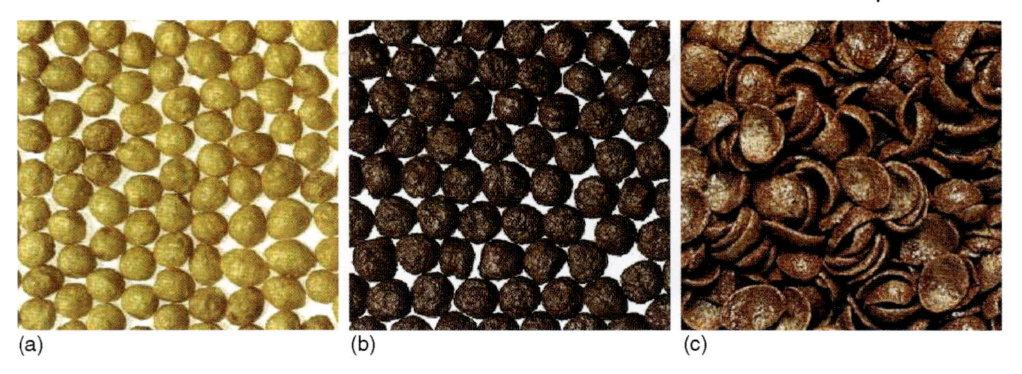

Figure 4.7 Different samples of direct expanded coated breakfast cereals: (a) sugar coated, (b) and (c) chocolate coated.

forming extrusion. This permitted the use of a wider range of feed materials which might not otherwise readily form into grits suitable for flaking and virtually eliminated the fines recycling [1, 4, 8, 11]. Nowadays, newcomers, and even global players, are changing more and more to the extrusion technology.

In Figures 4.9–4.12 are shown typical production set-ups, both traditional and semi-traditional (implementation of extrusion forming) for production of cornflakes with samples.

In recent years the production of co-extruded products has become very popular. Pillow-shaped extrudates with a tasty filling (Figure 4.13) have proved to be a very attractive foodstuff, especially appreciated by children. They are finally formed in an additional set of equipment consisting of a specially designed die, filling station and cutting device [4, 10, 11]. During production, the extrudate leaving the extruder in a tube form is simultaneously filled with cream then flattened between two rollers (crimping + cutting); after that, it is cut into pieces in such a way as to seal both open ends. This is done by the second pair of the rollers, where one is equipped with cutting tongues. The design of this equipment is shown in Figure 4.14. The cream is

Figure 4.8 Different samples of direct expanded breakfast cereals (extrudates produced with a multi-color or bi-color system on extrusion).

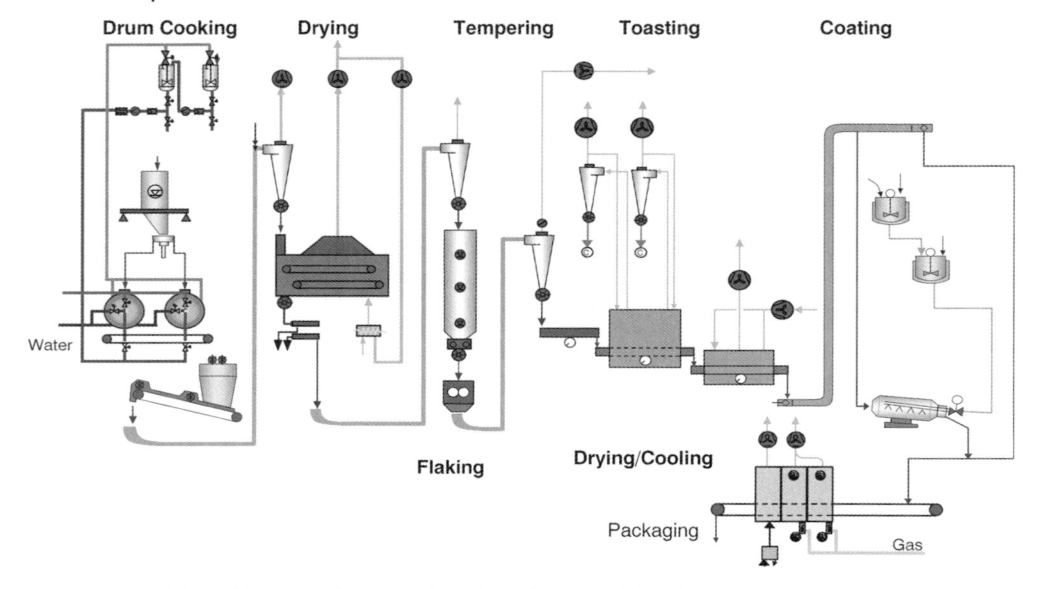

Figure 4.9 Example of a traditional "cooking" cornflakes line [13].

encapsulated by a hard hull (outer recipe – extruded for example, with a certain content of sugar). Finally, the pillows are dried and cooled. Generally, no coating is necessary or used.

The tubes are filled by means of special feeder devices which supply the mass to the head with a special pump synchronized with the screw rotation (Figures 4.15 and 4.16).

Extrusion-cooked cereal flakes. The production of extruded corn flakes or multigrain flakes (see Figure 4.17.) is performed on a much more complex technological line than in the case of directly extruded breakfast cereals. Certainly, such a line is much more expensive to acquire.

Multigrain flakes is the latest evolution in flakes. It is only possible to produce these by extrusion-cooking. Multigrain flakes look very healthy (see Figure 4.18).

Figure 4.10 Samples of traditional cornflakes (less uniform in shape and color).

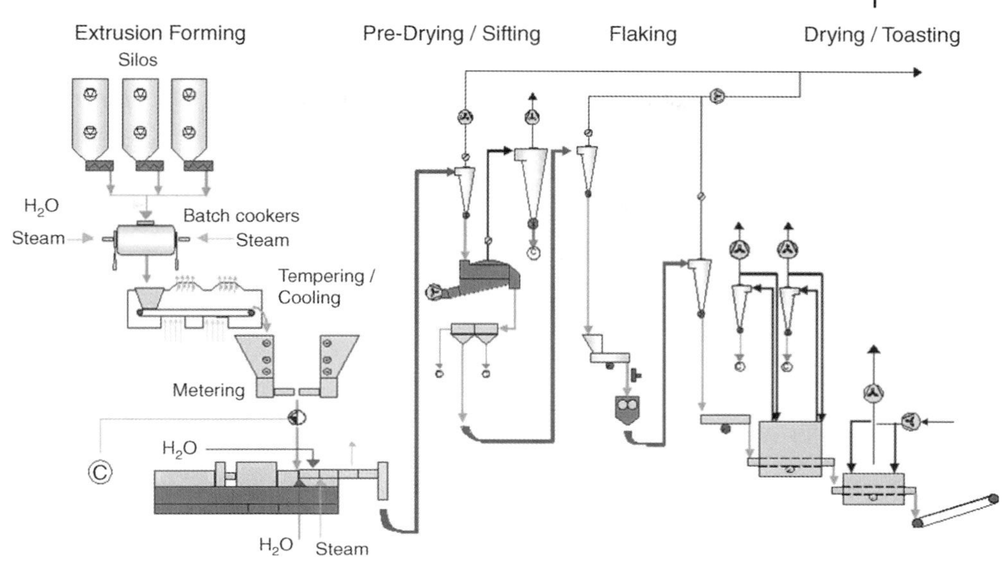

Figure 4.11 Example of a semi-traditional cornflakes line equipped with a twin screw extruder (here without any coating, it can be used in the same way as the traditional set-up) [13].

Now, a few words about the basic machinery and production equipment as well as the applied technology. In the past an extrusion-cooking process was usually performed in two extruders:

- The extruder – gelatinizer (known professionally as the G extruder) equipped with a large mixer-conditioner, in which the finely ground material (single or compound) after the addition of 20 to 30% of water, in relation to the dry substance, is subject to homogenization, mixing then baro-thermal processing at low shear and at a temperature below 160 °C. The main role of this extruder is to gelatinize the starch to obtain a "cooked" dough-like mass subsequently fed in portions to the next extruder;

Figure 4.12 Samples of semi-traditional cornflakes.

Figure 4.13 Samples of filled pillows.

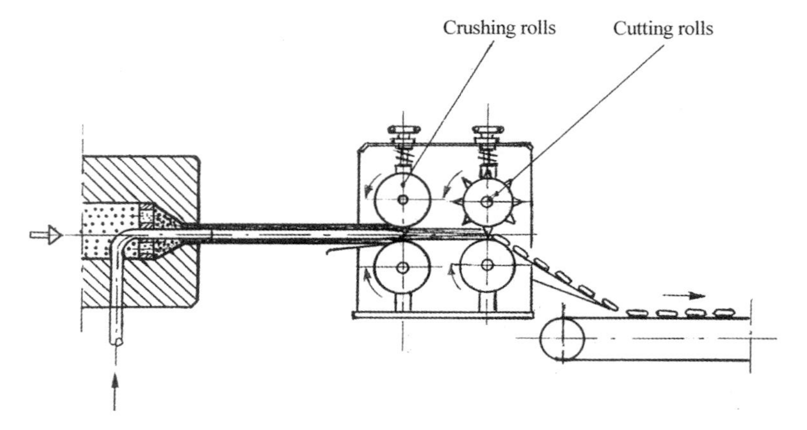

Figure 4.14 A cutter with one pair of rollers preparing co-extruded products.

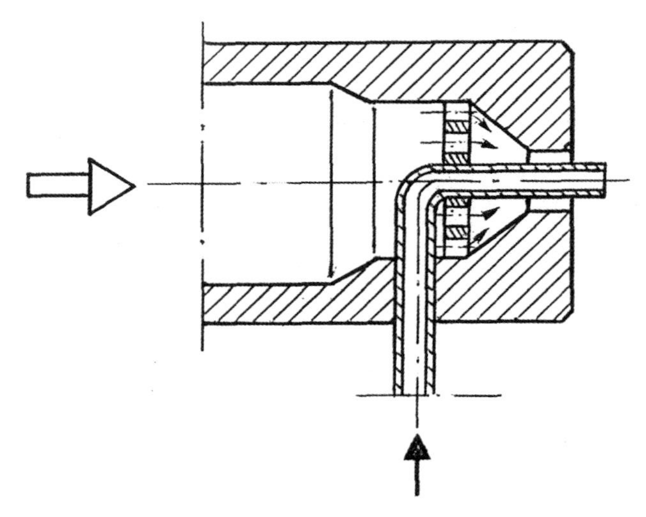

Figure 4.15 The filling of co-extrudates.

Figure 4.16 The filling station – a device for the preparation and forcing of mass filling the extrudates (permission of Bühler AG).

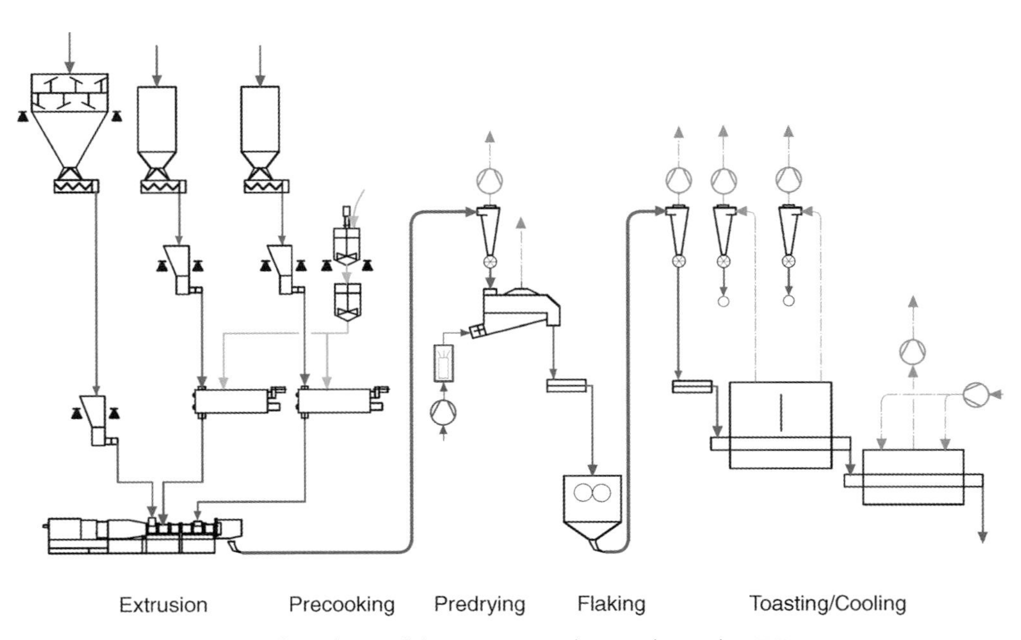

Figure 4.17 Typical set-up of a multigrain flakes extrusion-cooking production line [13].

Figure 4.18 Samples of extrusion-cooked multigrain flakes.

- The former (F extruder), which consolidates the mass density, eliminates gas bubbles and sets the ultimate shape of the products, most commonly as balls, by means of a high-speed knife. This process takes place at a much lower temperature, namely 60–70 °C.

Nowadays, the process is split into a hydro-thermal (by pre-conditioning) and a mechanical (extrusion) process with cooking and forming by twin-screw extruder. A typical extruder set-up for many applications, not only in the cereal sector, is shown in Figures 4.19 and 4.20, followed by an example of a preconditioning process diagram presented in Figure 4.21.

The next stage of production is flaking, preceded by pre-drying of balls of dough in order to prevent them from adhering. The formation of flakes takes place in a flaker equipped with two crushing rollers made of hardened, surface-polished steel which are constantly cooled and have a spacing gap mechanism (Figure 4.22).

The final shape and texture of the flakes are set in the next device – a toaster – at a temperature of 150–300 °C. Within a few dozen seconds, flakes are subjected to intensive drying in the hot air, most often on a perforated tape moving beneath a

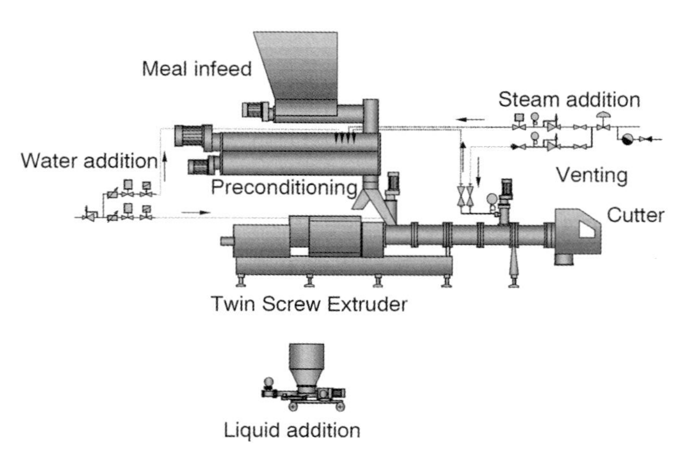

Figure 4.19 Principle of the twin-screw extruder set-up for processing of cereals [13].

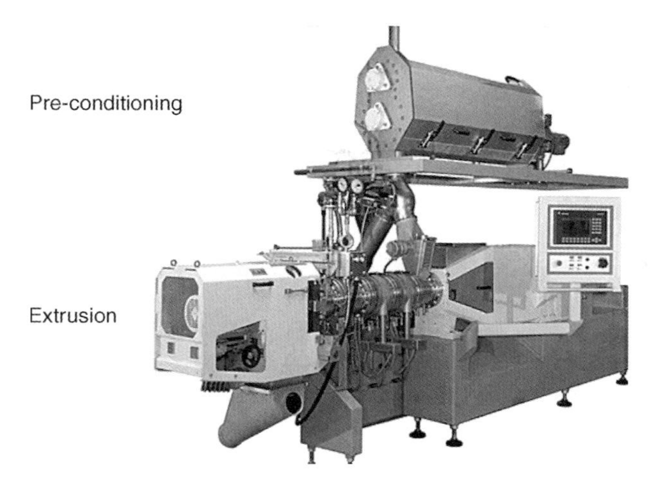

Figure 4.20 Example of a twin screw extruder set-up for cereals (permission of Buhler AG).

number of air nozzles and passing through subsequent sections of the device. By adjusting the equipment settings, the operator may control the processes of flake toasting and coloring in order to obtain the desired crunchiness. Modern toasters and driers for cereals are single pass up to multi-pass driers/toasters with a dual plenum air system for the highest uniformity (see Figures 4.23 and 4.24).

The flakes after cooling are conveyed to packaging or undergo further upgrading treatment, that is, spraying with flavor additives with micronutrients, sugar coating, sprinkling with cinnamon, peanuts or mixing with dried fruit. Thermophilic vitamin additives are applied in the last stage before packing.

From the technological point of view, the most demanding process is the application of sugar coating. In industrial production, in order to make a high quality product, a high-performance dryer with a cooler is then required. Flakes can easily adhere or absorb humidity which, after having been packed, may result not only in lowered quality but also may pose a threat to health.

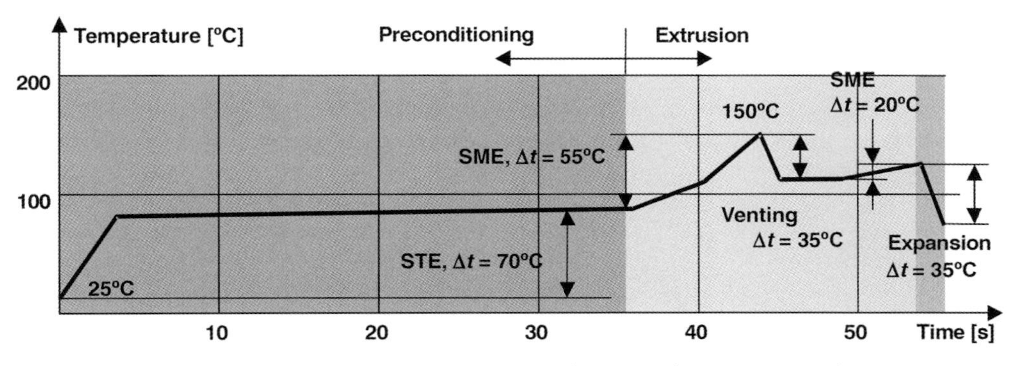

Figure 4.21 Process time and temperature diagram in pre-conditioning with twin-screw extruder.

Figure 4.22 A flaker (permission of Buhler AG).

Figure 4.23 Three-pass cereal drier (permission of Buhler-Aeroglide).

Figure 4.24 Single-pass toaster/drier (permission of Buhler-Aeroglide).

A modern technological line, whose main component is, for example, a multi-functional twin-screw extruder, can be assembled from a number of "mobile" auxiliary machines; this allows the manufacture of not only extruded and toasted cereal flakes, but also of many flavours of crisps, pre-cooked instant pasta, or the so-called pellets (a semi-finished product intended for frying and purchased by snack manufacturers).

4.4 Remarks on Operation

Initially, the production of extrusion-cooked cereal flakes was performed on regular high-pressure single-screw extruders with slightly modified plasticization systems reducing shear stress in the processed material. Soon after, low-shear extruders appeared to produce the best results. What is more, the improvement of dough gelatinization was achieved through the application of steam. Since the effectiveness of the new techniques on single-screw extruders was below expectations, the process was done before the proper extrusion in special, integrated conditioners, where the mixture was promptly subject to hydration and heat treatment. Such a solution is currently preferred by many extruder producers.

The construction of co-rotating twin-screw extruders at the end of the 1970s dramatically aided operators and technicians in the proper thermoplastic handling of dough before flaking. This was possible owing to the adjustability of the screws' geometry, especially in extruders of modular design. Modern extruders are able to perfectly knead and plasticize dough in low pressure conditions and with limited impact of shear stresses.

In recent years, leading manufacturers of extrusion-cookers have begun to offer modified twin-screw extruders used for the production of RTE cereal flakes. They proposed plasticization units extended by additional modules serving as formers. This helped to eliminate formers as an extra device in the technological line.

By pre-conditioning with a co-rotating twin screw extruder the process is split into three parts:

$$SME\ [kWh/t] = \frac{2 \cdot \Pi}{60\ [s/\min]} \cdot \frac{screw\ speed\ [1/\min] \cdot torque\ [kNm]}{throughput\ [t/h]}$$

$$STE[kWh/t] = m \cdot \Delta t \cdot c = \frac{\Delta t \cdot ((m_s \cdot c_s) + ((m_w + m_{wa}) \cdot c_w))}{100} \cdot \frac{1000[kg/t]}{3600[s/h]} = \frac{[kJ/kg]}{3.6}$$

Figure 4.25 Pre-conditioning with co-rotating twin-screw extruder.

- The pre-conditioner to provide the hydro-thermal energy in the product by steam and water addition (see Figure 4.25 ⇒ STE). Its job is also a very good mixing and to get a high dwell time.
- Extrusion by cooking in the first section and degassing/forming in the second by mechanical energy, as shown in Figure 4.25 (SME). For indirect expanded products forming consists only of flaking/toasting.

The extruders are characterized by L/D equal to or greater than 12 for easy applications like direct expanded products, and up to 20 or 24 for example, for indirect expanded products like cornflakes or multigrain flakes.

When it comes to the extrusion-cooking of cereal flakes, the more time taken for heat treatment, the better the texture and taste of the gelatinized dough leaving the extruder. This, however, entails an increased energy consumption and excessively generated heat in the cylinder, which is not desirable. The solution is a compromise, that is, the proper selection of equipment and optimization of processing.

However, there is also another factor limiting the effectiveness of the extruder's operation, namely its output performance. As regards high-performance devices, that is, offering an output of more than 1200–1500 kg of extrudate per hour, the applied method is extrusion of gelatinized dough in the form of a bunch of single strips (similar to thick pasta), instead of pellets cut directly at the head into the shapes of balls or ellipses. Of course, in such a case, the process of cutting the dough before flaking is carried out differently.

Despite the many advantages of the use of dual-function extruders, still many producers of breakfast cereals prefer to use two separate devices: a G-type extruder and a former. There are several reasons for this. The most important are: greater freedom in terms of produced assortment, no need for degassing of the dough and greater cooling efficiency before and during forming.

A standard former is a relatively inexpensive device equipped with a short single screw plasticizing system ($L/D < 8$) and consisting of a furrowed barrel and "deep" screw of a conveying type (the compression ratio from 1.5 to 3). The device has a water jacket; the barrel cooling system ensures a stable temperature of the dough during extrusion within 60–90 °C. Gelatinized dough is fed into the barrel in a way imposed by the screw feeder. The process of forming takes place in the head while the dough is being forced through a multi-hole die; then, it is cut by a rotary knife adjacent to the surface of the matrix face.

At this point, the importance of resistance while forcing the mass through the die should be stressed. The dies should be small, their number suitable and their channels relatively short. This will help to avoid the undesirable phenomenon of expanding pellets, whose maximum expansion level cannot exceed 1.5. A large number of holes, and thus the amount of cut pellets, may contribute to their adhesion and the formation of agglomerates. This can be alleviated by the application of cool air at the head.

As already mentioned, some manufacturers use a different method, namely form dough as bands or thicker pasta which are cut on a special drum cutter at a certain distance from the head. In the meantime, they are already sufficiently cool.

Modern formers can also be used for the production of potato pellets used for toasting and/or fried potato chips. They must be equipped additionally in the material feeder and mixing conditioner, as shown in Figure 4.26. In this case the technological line does not require a G–extruder, because the dough plasticization process is relatively easy and is usually performed at a temperature between 55 and 70 °C [12].

The optimum moisture content of flake pellets, typically extruded in the form of small balls or eights, should range between 15 and 24% depending on the raw materials used. Dough moisture content determines the susceptibility of pellets to flaking (friction between flaker's rollers), which ultimately affects the texture of the flakes. The flaking process is performed on special flakers, and is of significance because it determines the quality of the final product. It can be adjusted in multiple ways. For example, in order to obtain harder flakes, drier pellets are needed. In order to acquire a more crunchy and bubble-like structure after toasting, the pellets must be moisturized and surface-dried.

During the toasting of flakes in a toaster at a temperature from 150 to 300 °C, they acquire the ultimate shape and quality characteristics. If appropriately adjusting the time and temperature of toasting, it is possible to influence the crunchiness, color and flavor. The texture of the flakes depends largely on the degree of gelatinization of the dough, their thickness and moisture content.

Figure 4.26 Multi-purpose single-screw former (permission of the Pavan Group).

Since the extrusion technique offers much greater opportunity for influencing the physical properties of semi-finished products, for example, through greater diversity in the methods of preparation of dough in the initial production stage, the range of breakfast cereals available on the market has greatly increased. Notwithstanding, there is still much to be explored in this field.

References

1 Caldwell, E.F., Dahl, M., Fast, R.B., and Seibert, S.E. (1990) Hot cereals, in *Breakfast Cereals and How They Are Made* (eds R.B. Fast and E.F. Caldwell), American Association of Cereal Chemists, St. Paul, MN.

2 Fast, R.B. (1987) Continuous Process For Cooking Cereal Grains, U.S. Patent 4, 699, 797.

3 Fast, R.B. (1990) Manufacturing technology of ready-to-eat cereals, in *Breakfast Cereals And How They Are Made* (eds R.B. Fast and E.F. Caldwell), American Association of Cereal Chemists, St. Paul, MN.

4 Frame, N.D. (ed.) (1994) *The Technology of Extrusion Cooking*, Blackie Academic and Professional, London.

5 Guy, R. (2001) *Extrusion Cooking, Technologies and Applications*, CRC Press Inc., Boca Ration, FL.

6 Harper, J.M. (1981) *Extrusion of Foods*, CRC Press, Boca Raton, FL.

7 Hoseney, R.C. (1994) *Principles of Cereals. Science and Technology*, American Association of Cereal Chemists Inc., St. Paul, Minnesota, USA.

8 Mercier, C., Linko, P., and Harper, J.M. (1989) *Extrusion Cooking*, American Association of Cereal Chemists Inc., St. Paul, Minnesota, USA.

9 Miller, R.C. (1988) Continuous cooking of breakfast cereals. *Cereals Foods World (J. CFW)*, **33** (3), 284–291.

10 Miller, R.C. (1991) Die and cutter design (lecture notes), in Food Extrusion, American of Cereal Chemists Short Course, June 17–19, Leuven, Belgium.

11 Mościcki, L., Mitrus, M., and Wojtowicz, A. (2007) *Technika ekstruzji w przemyśle rolno-spożywczym (In Polish)*, PWRiL, Warszawa.

12 www.pavan.com (2010) Information & technical materials.

13 www.buhlergroup.com (2010) Information & technical materials.

5
Snack Pellets

Leszek Mościcki

5.1
Introduction

In recent years, in addition to extrusion-cooked corn flakes and breakfast cereals, a new generation of expanded products has become exceedingly popular: fried (sometimes toasted in hot air) cereal and potato snacks. They are produced in two stages: the output of the first stage is the so-called pellets (mostly manufactured by specialized firms), which, in the second stage, are fried by the producers of the final snack foodstuff [1–7]. These producers also enrich them by applying flavors and, finally, handle packaging and delivery to retailers.

Pellets are extrusion-cooked semi-finished products manufactured from a variety of mixtures of starchy materials, mainly wheat flour, maize flour and processed potato. They can be easily kept in dry storage for 12 months. Their bulk density is between 0.3 and 0.4 $g\,cm^{-3}$ (10 times greater than the density of ready-to-eat products) and the humidity does not usually exceed 8%. These products are prepared for consumption in two ways. Most often they are fried in deep oil in special frying devices or toasted by the application of hot air (see Figure 5.1). They can also be fried in a pan at home. In principle, this should not be done because there is no guarantee of obtaining good-quality snacks (because of the irregular frying temperature).

As already mentioned, the production of snack pellets relies upon a number of different starch materials, mainly cereals and potatoes. The total starch content in the recipe should be 60% at a minimum in order to obtain snacks with a crunchy texture [8, 9]. Maize pellets made of maize flour of different degrees of fragmentation may be an example. During toasting, maize pellets expand easily being light and rustling.

Table 5.1 gives examples of different recipes of mixtures used in the USA for the production of cereal pellets of varied texture.

With the high content of tapioca starch or potato starch in the recipe, there may be some factors that impede the production of pellets. This is related to the low temperature of starch gelatinization, its high viscosity, strong water absorption and specific bonding properties.

Extrusion-Cooking Techniques: Applications, Theory and Sustainability. Edited by Leszek Moscicki

ISBN: 978-3-527-32888-8

Figure 5.1 Various forms of snack pellets and expanded snacks.

The application of a modern twin-screw food extruder enables efficient processing of raw materials of starch content ranging from 70 to 100%. In such circumstances, single-screw food extruders may cause significant operational problems [5, 6].

Tapioca flour is the basic ingredient of mixtures used in the manufacture of pellets in Asian countries. Considering the functional characteristics of starch, such as the size of the granules, the low gelatinization temperature and amylose chains five times longer than in cereal grains, it has become a popular material in that region of the world. The total lack, or low content of fat and protein, good bonding properties, white

Table 5.1 Examples of mixture composition for cereal pellets used in the USA [6].

No.	Material	Frail pellets		Hard pellets	
		Recipe 1 [%]	Recipe 2 [%]	Recipe 1 [%]	Recipe 2 [%]
1	Wheat flour	—	7.0	—	94.5
2	Maize flour	56.0	—	94.5	—
3	Rice flour	—	9.0	—	—
4	Tapioca flour	—	80.0	—	—
5	Wheat starch	27.5	—	—	5.0
6	Maize starch	—	—	5.0	—
7	Tapioca starch	14.0	—	—	—
8	Vegetable oil	2.5	3.0	—	—
9	Salt	—	1.0	—	—
10	Monoglycerides	—	—	0.5	0.5

color and sweet flavor are some other advantages of tapioca recognized by their snack producers. In Europe the most widely used materials are: potato starch (basic ingredient of potato pellets), ground dried potatoes (adding characteristic flavor) and potato flakes (a structure-forming factor improving the stretch of the dough).

In the production of pellets, various types of additives can be included in the recipe. The most common are:

- thermally or chemically modified starch to improve the rheological properties of the dough;
- fragmented protein products of animal origin (shrimps, fish, poultry, beef, cheese, powdered milk) and of vegetable origin (seeds of leguminous plants), which are added in quantities up to 30% and raise the nutritional value of the end products.

5.2 Methods of Snack Pellet Manufacturing

Depending on the nature of the equipment installed on the technological line, different types of pellets can be obtained:

- short forms – cut off directly at the die of the extruder (similar to short pasta);
- laminated and/or perforated – cut out of a sheet of dough;
- 3D (three-dimensional) – produced from two sheets of dough adhering to each other at the edges during cutting.

5.2.1 Production Stages

5.2.1.1 Initial Phase

After preparation and homogenization of the raw materials mixture, it is pneumatically transported to the extrusion unit, where it is subjected to conditioning. The primary purpose of conditioning, besides the appropriate mixing of materials with liquid components, pigments, flavours or additives, is hydrothermal treatment and preliminary gelatinization of starchy components. From the conditioner the material is directly fed to the extrusion-cooker for baro-thermal processing. In the first stage, the mass is finally gelatinized in order to be formed in the second.

5.2.1.2 Extrusion and Forming

During extrusion-cooking, the processed materials undergo significant physical and chemical changes. By appropriate selection of process parameters such as temperature, moisture content and processing time, the mass becomes entirely gelatinized and cooked. Depending on the nature and content of starch in the mixture, extrusion-cooking parameters are varied. In the case of cereal pellets, the working temperature range in the gel unit of the extruder is set between 90 and 150 °C and in the formation unit between 65 and 90 °C. During the production of potato pellets, extrusion can be practically limited to the forming activity and no two-step baro-thermal treatment is required. Moisture content in the mixture in both cases should range from 25 to 35% and the extrusion time from 30 to 60 s.

The sources of heat are: steam fed directly to the mixture, heating units of the barrel and the transition of mechanical energy into heat. All of these are carefully monitored and adjusted by the appropriate automatic controls of the extruder. The control and monitoring of the production process by means of computer systems fitted with suitable software expedites the maintenance of high stability and efficiency of equipment operation during the production of pellets.

The gelatinized mass of dough is forced to the last forming unit of the extruder (prolonged barrel), where it is cooled and shaped, depending on the needs, in sheets (longitudinal shapes cut in further processing) or cut like pasta (short forms).

In former solutions, as in the case of the extrusion of cereal flakes, the bits of gelatinized dough were transported to a separate device – an F- type former.

The selection of suitable dies has an enormous impact not only on the shape of snack pellets but also on their functional characteristics and the quality of the finished product. The latest double-function twin-screw extrusion-cookers are gradually replacing the previously used ones: G and F (see Figure 5.2). They prove especially effective in the manufacture of a fixed range of pellets and are more convenient and less expensive to maintain [7].

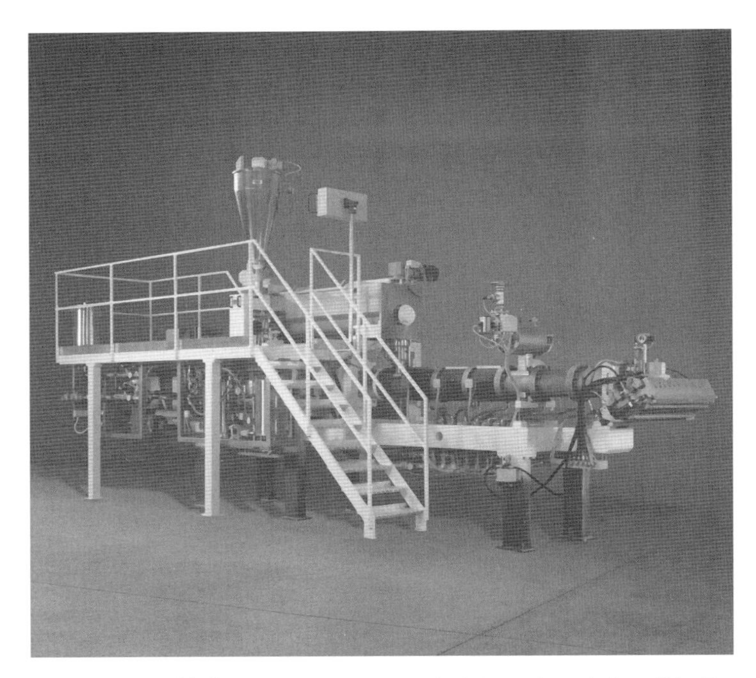

Figure 5.2 Double-function twin screw extruder TT-type (permission of the Pavan Group).

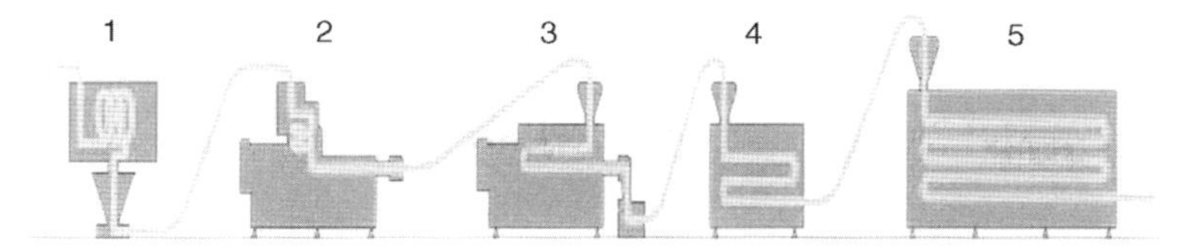

Figure 5.3 A production line for short form cereal snack pellets: 1 – mixer, 2 – extruder, 3 – former, 4 – pre dryer, 5 – dryer [7].

5.2.1.3 **Production of Short Forms**

Gelatinized and pre-cooked dough is transported to the former (or cooling forming section of the extrusion-cooker), where it is forced through the device dies (Figure 5.3). The final form and longitudinal size of the pellets depends on the shape of the die holes and the rotation of the knife mounted directly on the extruder's head. Most frequently, the produced snack pellets have very sophisticated shapes, for example, resembling fish, stars or bicycle wheels. The formed products are transported to a dryer for initial tempering and drying of the external layer (in order to prevent adhering) and then put to the final multi-stage drying process lasting until the desired product moisture content has been reached. Depending on the recipe, final products should contain from 6 to 10% of water. This ensures the maximum degree of expansion during further treatment.

As already mentioned, extruders are not involved in the manufacture of potato pellets in the way that formers are. The following heat treatment is limited to the temperature range 50–80 °C.

5.2.1.4 **Production of Laminated, Perforated and Spatial Forms**

Laminated pellets are made from wide sheets of dough pre-rolled on special devices which set their shape and uniform thickness (Figure 5.4). Extrusion-cooked dough should be free from air bubbles and display flexibility and rheological homogeneity. The point is that the process of cutting pellets runs properly and the sheet is not damaged or broken. The product must have stable quality – it will be put to a harsh test during the toasting or frying of snacks.

Perforated pellets are made from sheets of dough with a high degree of starch gelatinization, similarly to the laminated pellets. The difference lies in the fact that the roller-cutter unit has a spiked surface which allows perforation of a sheet of dough while maintaining its uniform thickness. Perforated pellets are ranked among the most difficult to make because of the high susceptibility to damage. Due to a large

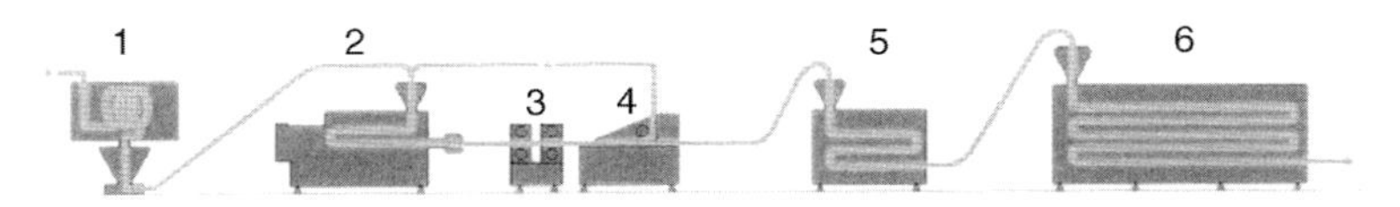

Figure 5.4 A production line for laminated potato snack pellets: 1 – mixer, 2 – former, 3 – lamination unit, 4 - cutter/slicer, 5 – pre dryer, 6 – dryer [7].

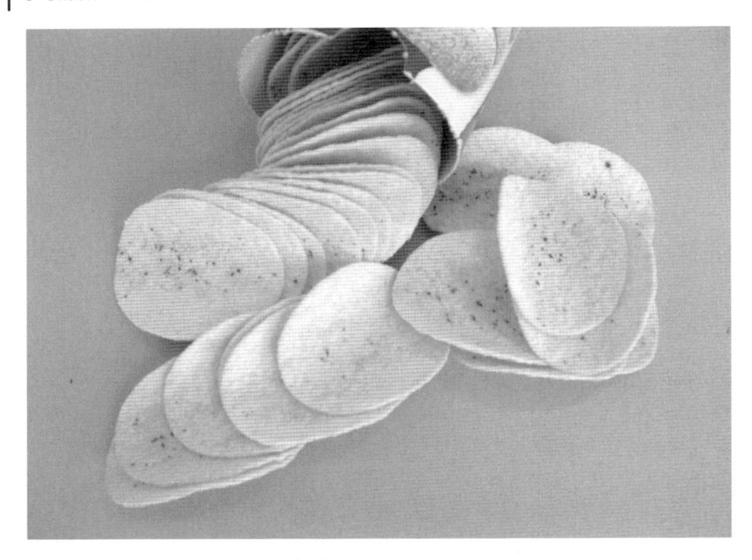

Figure 5.5 Potato snacks packed in tubes.

surface for the exchange of heat and mass during roasting, they produce remarkable results in terms of their gain in volume.

For several years, "luxury" potato snacks packed in a characteristic tube have gained considerable popularity (Figure 5.5). The product has been pioneered by Pringles®. These snacks are cut from a band of dough made in an F-type extruder and shortly after that roasted or toasted directly on the line [6, 7]. During toasting they are so shaped as to resemble the letter C. After cooling, they are vacuum-packed on special packing machines. Such a production line is functional, has a compact design and can be entirely automated.

A recent market hit have been spatial snacks made from a new generation of snack pellets called "3D", which resemble three-dimensional geometric shapes (see Figure 5.1). They are produced on an extrusion-cooker making two wider sheets which, after initial drying, are grouped together in special roller-cutter machines (see Figures 5.6 and 5.7). The edges of the pellets are stuck together and cut out from the double sheet by special roller-stampers, just as we do when preparing simple ravioli at home. Steel, Teflon-coated stampers are among the most expensive components of the auxiliary machinery used in technological lines. The products obtained are various shapes of pellets, including animals, hearts, pads, figurines and other spatial forms.

Quite recently, 3D-3L pellets have been developed and patented. They are made from three perforated layers, with the centermost layer having a distinct flavor and color. They are an interesting alternative to single-flavor snacks.

The production of spatial pellets is undoubtedly the most demanding field of extrusion-cooked food production. It requires expensive, specialized equipment, much attention and care in its handling as well as superbly trained production personnel.

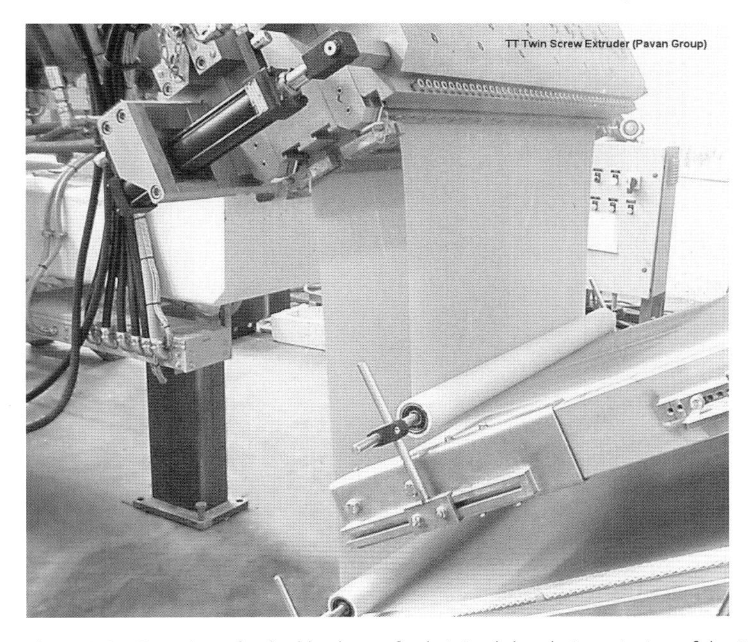

Figure 5.6 Extrusion of a double sheet of gelatinized dough (permission of the Pavan Group).

5.2.1.5 **Drying, Packing and Storage**

In the first phase of production, pellets contain from 20 to 30% water; therefore, before packing and storage, they must be dried to a moisture content level below 10%. A drying phase plays a very important role because this is when the products receive the definitive quality characteristics, that is, adequate stability during storage, susceptibility to expansion and durability. They must have the so-called glassy appearance be free from air bubbles, cracks and fissures. Furthermore, they should display a balanced distribution of humidity in the entire mass and have a smooth and non-deformed surface.

The process of drying snack pellets is conducted in two stages: preliminary drying, and final drying. Before shipment to the recipient, the pellets must also go through a tempering phase for a minimum of 70 hours to obtain an even distribution of humidity in the product. The most commonly used drying devices are belt dryers,

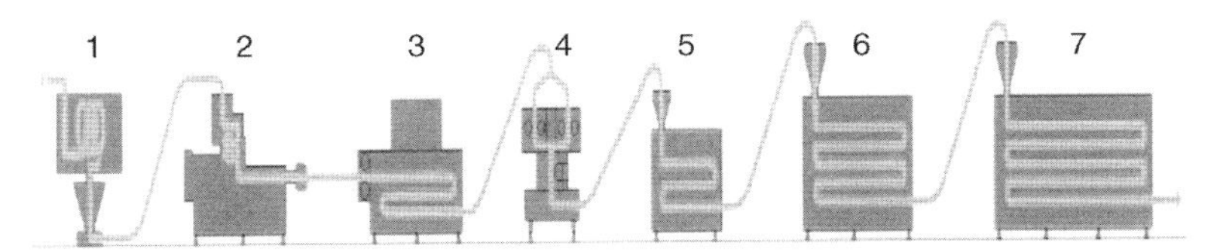

Figure 5.7 A production line for 3D cereal snack pellets: 1 – mixer, 2 – double function extruder, 3 – cooler, 4 - cutter/slicer, 5–7 – phase drying [7].

Figure 5.8 Pre dryer and multi-chamber drying unit for snack pellets (permission of the Pavan Group).

and, to a lesser extent, drum dryers. Dryers maintain a required temperature and humidity and the product residence time can be freely adjusted.

The most popular belt dryers for snack pellets are composed of several separate belts passing through chambers, so that the product is subjected to various stages of drying in varied climatic conditions (Figure 5.8). Pellets stacked in layers on perforated bands are dried by hot air while passing from one chamber (zone) to another, which takes from 4 to 8 hours, depending on the assortment.

Dried pellets are transported by a belt conveyor to containers, usually 500 kg big-bags or packed in multilayer paper bags. During packing, the pellets' temperature drops by a few degrees. Very often fans are installed under the conveyor in order to prevent the formation of condensate on the surface of the pellets. A perforated pipe is often placed inside the containers to act as a channel facilitating the exchange of heat and moisture during the many hours of conditioning. This also prevents the development of undesirable microflora. After a period of tempering, the big bags are loosely tied and placed on pallets in storerooms to await distribution. The recipient of the pellets hangs the bags directly over the belt conveyor of a toaster/fryer feeder.

The process of drying snack pellets in continually operating belt dryers can last for 8 hours while in drying units of periodic operation it often requires a 15-hour cycle. The drying temperature should be maintained below 90 °C with relatively high humidity. This prevents the formation of cracks and cavities inside the pellets.

Potato pellets stored in optimal conditions, that is, in a dry and dark place, at temperatures from 15 to 25 °C and air humidity from 40 to 70% exhibit durability of about 12 months. In the case of cereal pellets, this time can be twice as long.

5.2.1.6 Toasting or Frying – Final Stage of Snack Production

As already mentioned at the outset, the final stage of snack production is frying or toasting and the application of flavors and sprinkle (as in the case of directly extruded snacks). Pellets containing tapioca starch or potato starch expand very well in hot air in toasters and in an electromagnetic field (microwave dryer). The best results are obtained during frying of snack pellets in deep oil, which is usually a mixture of specially selected vegetable oils and the so-called hard fat. Unfortunately, it involves fat absorption (up to 30%) which is not very welcome by dieticians, even if very tasty [10]. Recently, producers have promoted fat-free snacks which are expanded by alternative methods discussed above. This provides much more favorable effects in terms of health but with a loss in sensory quality. Not surprisingly, fat-free snacks do not enjoy much interest.

References

1 Fazzolare, R.D., Szwerc, J.A., van Lengerich, B., and Leschke, R.J. (1992) Extruded Starch Snack Foods, U.S. Patent, 5, 104, 673.
2 Frame, N.D. (ed.) (1994) *The Technology of Extrusion Cooking*, Blackie Academic and Professional, London.
3 Guy, R. (2001) *Extrusion Cooking. Technologies and Applications*, CRC Press Inc., Boca Ration, FL.
4 Huber, G.R. and Rokey, G.J. (1990) Extruded snacks, in *Snack Food* (ed. R.G. Booth), AVI Book, Van Nostrand Reinhold, New York, pp. 107–137.
5 Mercier, C., Linko, P., and Harper, J.M. (eds) (1989) *Extrusion Cooking*, American Association of Cereal Chemists Inc., St. Paul, Minnesota, USA.
6 Mościcki, L., Mitrus, M. and Wojtowicz, A. (2007) *Technika ekstruzji w przemyśle rolno-spożywczym (in Polish)*, PWRiL, Warszawa.
7 www.pavan.com (2010) Information materials.
8 Keller, L.C. (1989) Method for Producing Expanded, Farinaceous Food Product, US Patent, 4, 869, 911.
9 Lee, W.E. III, Bangel, J.M., White, R.L., and Bruno, D.J. (1987) Process for Making a Corn Chip with Potato Chip Texture, US Patent 4, 645, 679.
10 Trela, A. and Mościcki, L. (2006) Wpływ receptury oraz warunków ekstruzji na zdolność pochłaniania tłuszczu pelletów ziemniaczanych. *Przegląd Zbożowo-Młynarski (in Polish)*, 8, 29–31.

6
Crispbread, Bread Crumbs and Baby Food

Leszek Mościcki

6.1
Production of Crispbread

Another example of the potential use of the extrusion-cooking technology is the production of crispbread from popular cereals. These types of products (often called extruded flat bread) have gained tremendous recognition worldwide, especially owing to their health values and long shelf life. Crispbread is an attractive form of bread, perfectly suited to sandwiches and snacks; when stored in a dry room, it has a few months' shell life (Figure 6.1). Its production method is twofold: conventional – baked in a tunnel bakery oven, and by extrusion-cooking, which is a novelty in this field [1–5].

To shed more light on the methods of production of extruded crispbread, let me use a specific example of such products manufactured in a handcraft sector facility in Poland [6, 7].

The main production unit is a twin-screw extruder (e.g., the one presented in Figure 6.2) being part of a simple processing line composed of a cutting device (the cutter), sorting table and packing machine, flow-pack type (see Figure 6.3).

The extruder has a die in the shape of a narrow slit which forms a ribbon-like shape of extrudate (depending on the performance and size of the extruder there may be a few of them), which is next fed into the cutting device to be finally rolled and cut to a desired length (flat plates).

A single-component material or a mixture of suitably ground cereal ingredients should be evenly fed directly into the extruder's barrel. The scope of the thermal and pressure treatment used during the production of crispbread is similar to that used during the production of directly expanded breakfast cereals. The operator's task is to control the work of the heaters, cooling and driving system by maintaining an appropriate temperature along the extruder barrel determined by the required technological regime.

Table 6.1 shows examples of a basic recipe recommended for the production of dietetic crispbread.

Extrusion-Cooking Techniques: Applications, Theory and Sustainability. Edited by Leszek Moscicki

ISBN: 978-3-527-32888-8

Figure 6.1 Samples of crispbreads – convenient and healthy products.

Figure 6.2 An example of a small twin-screw extrusion-cooker, type 2S-9/5 (Polish design); total installed power 40 kW, production capacity 150 –200 kg h^{-1}.

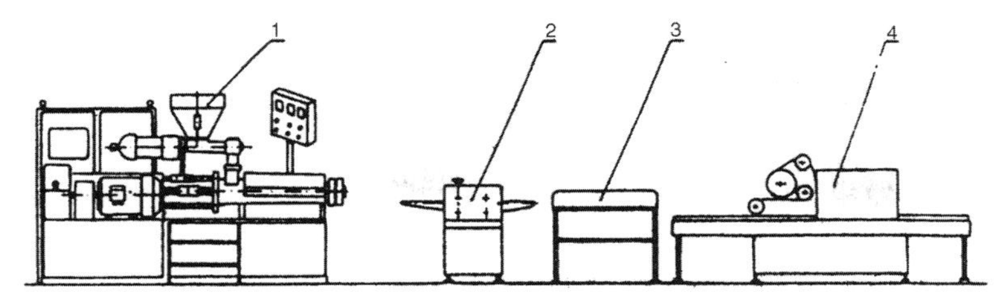

Figure 6.3 A diagram of a small processing line for the production of crispbread: 1 – twin-screw food extruder, 2 – cutting device, 3 – sorting table, 4 – flow-pack packing machine.

The finished product may be additionally toasted in the belt drier/toaster in order to enhance its sensory and organoleptic characteristics. It may also be entirely or partly coated with chocolate. Of course in such a case, the processing line should then be adequately extended by the necessary additional equipment. The final stage of production is packing in foil, trays, wrapping in foil or cardboard boxes.

Italian and British consumers have a liking for crispbread in the form of salty cereal sticks and/or cereal tubes filled with a sweet mass of high fiber content (the production of co-extruded products have been discussed in Chapter 4). Co-extrusion can also be suitable for production of cereal crispy bars made from extruded whole cereal grains filled with probiotics, which have become very popular nowadays.

6.2 Production of Bread Crumbs

Both traditional and contemporary cuisine requires a lot of products to be prepared by coating in various sorts of pannier or crumb. So far, the most popular material used at home has been and still is bread crumbs. It is used not only as a pannier but

Table 6.1 Examples of recipes for healthy crispbread [10].

Example I		Example II	
Material	**%**	**Material**	**%**
Rye flour	38	Wheat flour	60
Rye bran	25	Wheat bran	38
Pea flour	20	Salt and microelements	1
Maize grits	15	Herbs	1
Salt and microelements	1		
Herbs	1		

also as a filler in many meat dishes. In today's fast food era and the broad range of universal pre-prepared meat and fish products, requiring only some heating in a microwave or short frying, good quality panniers have attracted special attention. Popular bread crumbs of ground, stale bread of diverse origin do not seem to work well and satisfy the industrial food processing. It is not only the taste of the pannier that counts but also its texture and porosity, color uniformity and fat absorption. In the case of food production on automated processing lines, the stability of quality parameters of components is of paramount importance. Furthermore, much in demand are panniers rich in nutrients such as protein, fiber and micronutrients.

Bread crumbs may be produced with the extrusion-cooking technology. Owing to its universality, it is possible to adjust the physical and chemical properties of these products depending on the needs and their final purpose [8–10]. Extrusion-cooking enables the production of panniers from many cereal materials, including cereal bran. The addition of maize into a mixture of materials influences not only the color of the product (hot-yellow), but it also extends the physical features: increased porosity and water absorption as well as characteristic flavor. The same is true of rice, even wild rice (not dehulled). A small portion of potato flour in the recipe of the pannier helps obtain a golden-brown color of the coated meat during frying. The examples given clearly show that it is viable to manipulate the utility characteristics of the pannier, including additional treatment aimed to obtain color, taste and nutritional value.

When it comes to bread crumbs, the set of manufacturing machinery is very much the same as in the case of crispbread shown in Figure 6.1. It is even simpler. The main difference is the absence of a cutting device (the product is cut directly at the die with a fast-speed knife) and a sorting table. Instead, other devices are used, namely a rolling shredder or a hammer mill and screen sieves.

Screened and sorted product is packed in foil bags (addressed to retail) or larger collective packages (wholesale) by means of a vertical, volume packing machine, for example, Trans-Wrap type.

Strictly speaking, the industrial production of such products requires the use of much larger devices, joined by conveyors, where the total production process takes about 30 minutes. The electric power consumption of a processing line presented in Figure 6.4, performing with a capacity of circa $1.5\,t\,h^{-1}$ is about 300 kW [10]. The whole production process can be completely automated, which guarantees unchanging process parameters and a product of the desired quality.

6.3 Production of Precooked Flour, Instant Semolina and Baby Food

The majority of older readers will still remember preparing semolina with milk for their children and remember how much time and attention that baby food required to offer the intended effect. Today our stores offer a wide range of instant semolina and cereal foods for infants, their preparation consisting only in the appropriate

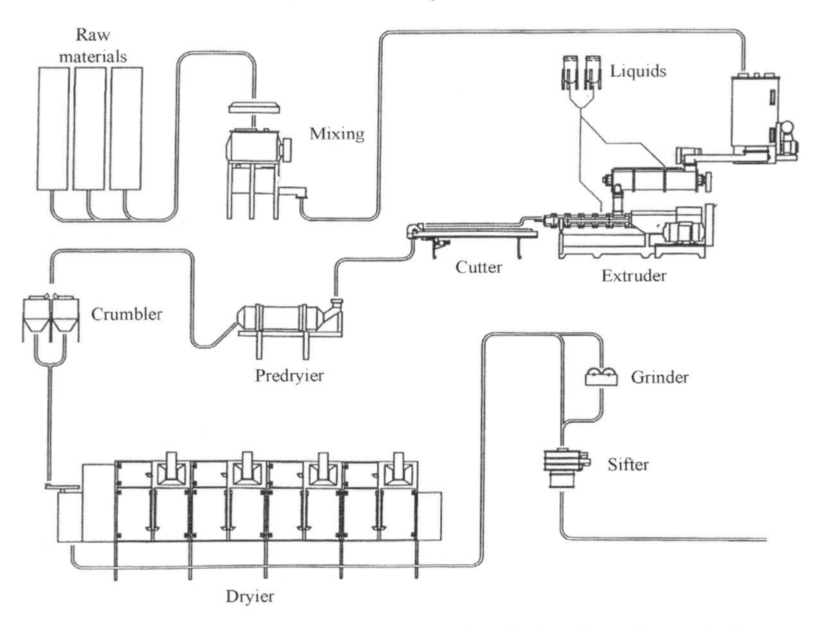

Figure 6.4 A diagram of an industrial processing line for bread crumbs production.

application of the contents of a package into hot milk or hot water. What is more, these products are rich in vitamins and micronutrients, may contain dried fruit and attractive additional flavours.

A breakthrough in the manufacture of the products discussed above has been the implementation of extrusion-cooking technology, which has largely streamlined the process of thermal treatment of raw materials. The compact design of processing machinery and total automation of their performance while eliminating the personnel's direct contact with raw materials and products enables the manufacturers to maintain a high level of hygiene and sanitary conditions. Today's production equipment is easy to use and offers a variety of applications (e.g., the production of various crisps and expanded extrudates). Their installation is relatively cheap, energy-efficient and does not require extensive maintenance staff.

A diagram of the relevant processing line for baby food is shown in Figure 6.5. After careful examination, it is immediately discernible that the line does not principally differ from those discussed in Section 6.2. The heart of the installation is, of course, a single-screw or better twin-screw food extruder (more popular due to greater versatility), which produces small cereal extrudates from rice flour, maize flour, wheat and/or buckwheat flour. They are subsequently dried to a humidity of 5% and very thoroughly ground. After being combined with other ingredients, such as milk powder, dried fruits and micronutrients, the product is mixed and packed [10, 11].

A comprehensive production facility for the multi-ingredient cereal baby foods performing at around 300–400 $kg\,h^{-1}$, presented in Figure 6.5, requires a production

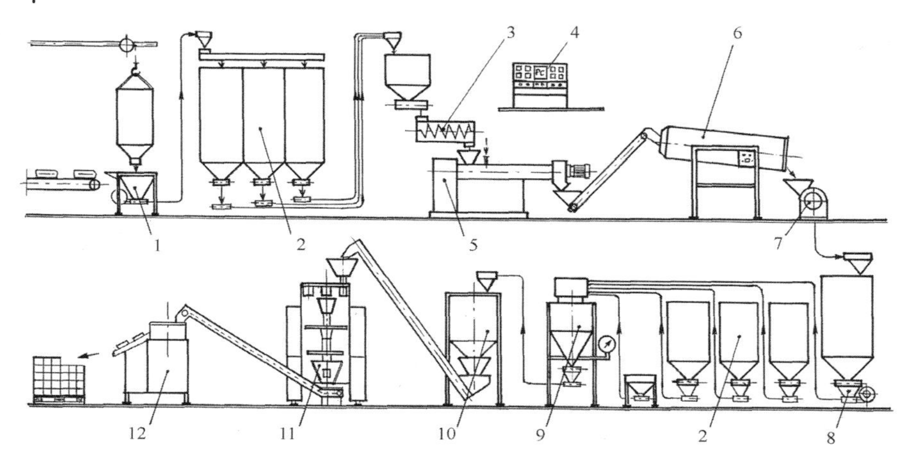

Figure 6.5 An installation for the manufacturing of precooked flour and multi-ingredient cereal baby food: 1 – unloading, 2 – silos, 3 – conditioner/mixer, 4 – control panel, 5 – extrusion-cooker, 6 – drum dryer, 7 – shredder/mill, 8 – pneumatic transporter, 9 – weigh, 10 – mixer, 11 – packing machine, 12 – palletizer.

area of 200 m^2 (including the storage of raw materials and product), 150–200 kW of electric energy and circa 6 $m^3 h^{-1}$ of natural gas [10]. With three-shift working, the annual production can reach a level of around 4.2 million 0.5 kg packages of product. The total cost of this particular processing line is about 1 million €. Of course, this can be done also on a smaller scale and with lower mechanization, which will considerably reduce the investment cost.

The production of baby food can be highly profitable and the incurred expenditure may be returned in a relatively short time. There is, however, one very important condition: the products of this type must be attractive, valuable and very carefully packed, and, above all, produced in appropriate health and sanitary conditions. These factors largely determine their success on the marketplace.

By properly adjusting the extruder's parameters, it is possible to produce extrudates of the desired density, tackiness, water absorption capacity, cold water solubility, and so on [1, 4, 5, 12]. This is particularly important in the so-called modified starch and pre-cooked cereal flour production, products used not only by the food sector but also by the pharmaceutical, paper and chemical industries. For example extrusion-cooked ground pea is a perfect ingredient for instant soup producers as an RTE type product. Nowadays, many other similar components of vegetable origin are used by ready-made or fast food producers. Modified potato starch has a large-scale application, since after extrusion-cooking it gains rheological features significantly different from those commonly known and related to a native starch. Recently it has been used effectively as a cooling agent in drilling installations on the North Sea. Cereal and leguminous extruded functional modifiers like Suprex® find extensive application in bakery, meat processing and feed industry [13].

References

1 Fast, R.B. and Caldwell, E.F. (eds) (2000) *Breakfast Cereals and How They are Made*, American Association of Cereal Chemists Inc., St. Paul, Minnesota, USA.

2 Frame, N.D. (ed.) (1993) *The Technology of Extrusion Cooking*, Blackie Academic and Professional, London.

3 Guy, R. (2001) *Extrusion cooking, Technologies and Applications*, CRC Press Inc., Boca Ration, FL.

4 Meuser, F. and van Lengerich, B. (1984) Possibilities of quality optimization of industrially extruded flat breads, in *Thermal Processing and Quality of Foods*, Elsevier Applied Science Publ., London, pp. 180–184.

5 Wiedmann, W. and Strobel, E. (1987) Processing and economic advantages of extrusion cooking in comparison with conventional processes in the food industry, in: *Extrusion Technology for the Food Industry* (ed. C. O'Connor), Elsevier Applied Science Publ., London, pp. 132–169.

6 Mościcki, L. (1991) Extrusion-cooking dietetic flat bread and a Polish example. I.C.C. Symposium, 10-13.06. Prague, Proceedings, p. 10.

7 Mościcki, L. (1991) Polish dietetic flat bread. 8th World Congress of Food Science and Technology, Toronto, Kanada, Proceedings, pp. 428–432.

8 Mercier, C., Linko, P., and Harper, J.M. (eds) (1989) *Extrusion Cooking*, American Association of Cereal Chemists Inc., St. Paul, Minnesota, USA.

9 Moore, D., Sanei, A., Van Hecke, E., and Bouvier, J.M. (1990) Effects of ingredients on physical/structural properties of extrudates. *J. Food Sci.*, **55** (5), 1383–1387, 1402.

10 Mościcki, L., Mitrus, M. and Wojtowicz, A (2007) *Technika ekstruzji w przetwórstwie rolno-spożywczym (In Polish)*, PWRiL, Warszawa.

11 www.fudex.com (2010) Technical information and materials.

12 Harper, J.M. (1981) Extrusion of Foods, vols. I and II, CRC Press, Boca Raton, Florida.

13 Mościcki, L. (2005) Suprex Food - funkcjonalne komponenty (In Polish). *Przegląd Zbożowo-Młynarski*, **5**, 38.

14 Abbott, P. (1989) Co-extrusion replaces complex filling systems. *Extrusion Commun.*, **2** (2), 11–13.

15 Hoseney, R.C. (1986) *Principles of Cereal Science and Technology*, American Association of Cereal Chemists, St. Paul, Minnesota.

16 Wiedmann, W. (1986) Suesse kochextrudierte Produkte: Verfahrenstechnische Grundlagen und wirtschaftliche Anwendung. *Suesswaren (Hamburg)*, **30**, 436–447.

7
Precooked Pasta

Agnieszka Wójtowicz

7.1
Introduction

The universal character of the extrusion-cooking technique provides the option of modifying the extruder by changing the configuration of the screws, the use of cooling or heating segments of the extruder and the application of various shape-forming dies. This has enabled the manufacture of not only directly expanded products, but also products intended for further processing.

Pasta is one of the most common sources of carbohydrate in a diet. Production and consumption of pasta products vary depending on the region of the world and culinary traditions within a society. In Poland the consumption of pasta is estimated at 5 kg per person per year, while an average Italian eats over 30 kg of pasta products such as spaghetti, lasagne, ravioli or special pasta every year [1]. The dynamic growth of small catering services and the increasing popularity of fast foods has stimulated the development of a new type of extrudates, that is instant pasta, that does not require cooking. Due to a thermal and pressure treatment, instant pasta is already precooked and requires only rehydration in boiling water or short cooking.

Precooked pasta is usually made following a conventional technology, supplemented with a pasta cooking stage in water or steam, or hot oil, followed by drying as in traditional pasta processing. The machinery set needed to produce pasta of good quality, proper rheological and sensory values consists of a mixer working at normal or low pressure, a press with a forming die, a drying unit set to a suitable drying cycle and a packaging device portioning the product into single packets [2]. In the case of extrusion-cooked pasta, the press is replaced by a single- or twin-screw extrusion-cooker [3–5].

There are many reasons for the introduction of the extrusion-cooking technique in pasta production. First, according to the thermal and pressure treatment, many raw materials can be used: wheat varieties, from soft wheat flour to semolina with various granulations [4, 6, 7], the so-called unconventional raw materials used usually as additives for pasta, such as legume flours [3, 5, 8–11] and starchy sources such as rice or maize [12].

Extrusion-Cooking Techniques: Applications, Theory and Sustainability. Edited by Leszek Moscicki

ISBN: 978-3-527-32888-8

Extrusion-cooking improves the universality of pasta products, particularly in terms of food preparation. Depending on the texture of the products obtained by appropriately selected parameters of the extrusion process, it is possible to offer not only traditional products, but also such prepared for a shortened final treatment, for example, in a microwave oven or in 2–3 minutes hydration in hot liquids (soups, sauces) [3, 4, 13, 14].

Precooked or instant pasta is a product of small diameter, which facilitates the process of hydration and preparation for consumption [5, 15]. During conventional processing of instant pasta, its feature of a precooked product with a high degree of starch gelatinization can be obtained during the hydrothermal treatment, that is, steaming, pre-cooking or frying [2]. These processes require special equipment and extra drying to reach the desired product storage humidity.

Processing in an extruder, the conditions for maintaining stabilization and customization of pasta are much simpler to meet and the drying time is shortened to 40–50 min at 70–80 °C. Consequently, production costs may be lowered compared with traditional methods. Another important advantage is the production simplicity, since the process does not require high-energy dryers, cooking in boiling water, steaming or deep oil frying [4, 10, 16].

7.2
Raw Materials Used in Pasta Processing

The most common raw material in pasta production is *Triticum durum* wheat semolina with granulation ranging from 200 to 300 μm and containing 0.8–0.9% of ash in dry mass, 12–13% of protein and no less than 30% of wet gluten (protein fraction determining the plasticity of dough) of good quality (gluten deliquescence of 8–13 mm), the falling number within 350–450 s, which provides high purity and quality of grains and a typical golden-amber color. However, due to specific climatic conditions this wheat variety is grown only in a limited area which reduces its availability and leads to high prices [2, 6].

As a substitute or supplemental material, manufacturers use farina or wheat flour from common soft wheat which, despite worse properties, after appropriate selection of processing parameters can make a good material for the production of pasta products. The soft wheat flour should contain no more than 0.4–0.5% of ash in dry mass, 10–11.5% of proteins with gluten content of 27–30% (gluten deliquescence 5–8 mm); it should originate from clean, not sprouted wheat (the falling number not less than 250 s) and maintain an adequate light-cream color [6, 17, 18]. Sample compositions of common flour types are presented in Table 7.1.

An important factor is also uniform granulation of the flour or farina from soft wheat needed for the correct hydration of flour molecules when mixed with water and additives. The flour used should be finely and uniformly ground because it affects the speed and equable hydration of pasta dough, as well as the final product. Water used for pasta dough preparation should not show high hardness and must conform to the parameters of drinking water [4, 19, 20].

Table 7.1 Composition of wheat products used in the production of extrusion-cooked pasta [1].

Feature	Flour type			
	450	500	550	Semolina
Wet gluten [%]	35	37	39	42
Dry gluten [%]	30	33	34	37
Gluten quality in Sadkiewicz units [S]	136	87	67	144
Sedimentation rate [ml]	22	24	23	32
Gluten deliquescence [mm]	5	7	11	5
Protein content [%]	11.02	11.98	11.78	12.42

In pasta production, various types of additives may be used such as eggs (in processing of egg pasta), vegetables (influencing the color and flavor of products), substances affecting the rheological characteristics of the dough, such as natural emulsifiers (soy or sunflower lecithin), or artificial chemical additives (mono- and diglyceride esters). Protein supplements are also added to improve the nutritional properties of the pasta or final product and enriching supplements, such as vitamins or minerals [2, 16, 21], are also added.

Pasta factories increasingly take advantage of common soft wheat flour or other alternative raw materials derived from cereals or legumes. These materials may be used as additives. Pasta based on traditional raw materials may be enriched with components with a higher content of nutrients, such as protein-rich sulfuric amino acids, aroma and flavor compounds and specific starchy ingredients needed to form the proper dough structure [2, 3, 17, 18, 22].

7.3 Extrusion-Cooked Pasta-Like Products

The technology developed by the Wenger Company (USA) shows the typical composition of extrusion-cooked pasta of firm consistency is: 98.0% semolina, 1.0% –monoglyceride, 1.0% powdered egg albumin. The soft texture of pasta may be obtained using only semolina (98.5%) and monoglycerides (1.5%) [13]. To improve the texture and reduce the hydration time it is necessary to change the addition of protein components, in particular, gluten, eggs, milk or emulsifiers [21]. The addition of L-ascorbic acid improves the formation of the protein matrix in pasta dough and reduces the loss of color during cooking [3, 16, 23]. On the other hand, the addition of methylcellulose reduces the adhesiveness of pasta made from soft wheat flour and can also influence the quantity of cooking losses [3, 16]. The production process may also involve the use of vegetable additives, such as carrots, spinach and pumpkin, or legume seeds, improving the taste and nutritional value of products [11].

The production of pasta by extrusion-cooking requires uniform granulation of raw materials, which prevents irregular water distribution on the starch granules during the mixing and kneading of the dough and an uneven surface of the pasta during extrusion [4, 6, 13].

An extremely important parameter in the extrusion process is the dough moisture content. When manufacturing simple pasta forms such as threads or spaghetti type, the dough moisture may be relatively low – around 28–29%, while for products with more complicated shapes, it is necessary to ensure higher dough moistening up to 30–32%. This allows smooth flow of the dough inside the barrel and spatial products juncture at the stitch [15]. Excessive dough moistening causes, especially when too short a screw is used, a significant decrease in pressure and accelerated dough flow which results in an unstable structure of the products after drying and a high amount of cooking loss [7, 9, 20, 24, 25]. An adequate pressure is determined not only by the dough nature but also by the correct efficiency ratio of the screw to the total surface of the forming die which can be adjusted by reducing the number of active openings, maintaining appropriate clearance between the screw and the inner, grooved barrel surface. This will prevent the dough turning with the screw if dough adhesiveness is too high (this applies in particular to single-screw extruders) [18, 24].

To avoid excessive stickiness, despite the additives used, pasta after extrusion may be subjected to a hot water bath followed by surface pre-drying and washing in order to remove starch collected on the surface [4, 8, 19]. After this treatment, it is necessary to dry the material at a temperature of around 90 °C in order to reach the moisture level of 12.5%. Starch in a processed product is approximately 90% gelatinized, the level of microbiological contamination is low and the ability for rapid rehydration is high, even in cold water.

7.3.1 Experimental Results

In the technology with a single-screw extruder application proposed by Wójtowicz [1], the operations of mixing, thermal processing and forming are performed in a single device, the starch is almost completely gelatinized and the drying time is relatively short. The manufacture of precooked pasta should be carried out in an extruder with a long plasticization unit and a die-head equipped with an intensive cooling system [1, 15]. During the processing of short pasta shapes, it is necessary to use a proper pasta die with rotary knife, the long spaghetti-type pasta may be formed in specific pockets [2, 4, 13]. Many research centers explore the techniques of producing instant or precooked pasta with a short preparation time but exhibiting the features desired by consumers, that is, appropriate color, taste, firmness and other qualities tailored to the traditions and tastes of consumers. Large manufacturers, like Pavan Group (Figure 7.1), have already extended their product range by instant products in the form of instant soups, desserts or dishes based on pasta products [2, 3, 9, 13]. However, the industry of ready meals has outrun science, so research on the influence of process parameters and the resultant physiochemical changes should be continued and expanded to include new materials or technologies. For several years, investigations has been carried out on the application of the extrusion-cooking technique in the production of precooked and instant pasta in the Department

Figure 7.1 A processing line for the manufacture of conventional pre-cooked pasta (permission of the Pavan Group).

of Food Process Engineering of the University of Life Sciences in Lublin, Poland [1, 7, 10–12, 14–16, 21, 23, 26].

When replacing pressing by the extrusion-cooking process, the amount of water remains at a similar level as in conventional technology, but there is a change in the range of process temperature in order to obtain a high level of gelatinized starch already in an extruder without the need for long drying at high temperatures [1, 15]. In our research program from 2000 to 2010, pasta was processed in the temperature range 70–100 °C using a modified single-screw TS-45 extruder with L/D = 16 or 18 (Figure 7.2), at different moisture levels of raw materials (28–36%) and varying screw speed in the range 60–120 rpm, connected with the performance of the device [1, 7, 10–12, 14–16, 21, 23, 26]. A forming die with 12 openings with a diameter of 0.8 mm was applied. After the processing, the pasta product was air-cooled to avoid stickiness of the pasta threads and subsequently dried for 1 h at 40 °C in a dryer with air circulation before finally storing in sealed plastic bags.

The intensity and extent of changes after thermal treatment may be defined by various food product quality determinants [2, 5, 9, 17, 24, 25]. When evaluating the pasta, a number of parameters are taken into account, including the minimum preparation time (cooking time), water absorption index, water solubility index, starch gelatinization degree, cooking weight, the amount of organic ingredients passing into the water during cooking (cooking loss), texture and pasta products sensory characteristics evaluated by instrumental methods (color, firmness, adhesiveness, viscoelasticity) and organoleptic methods (appearance, color, flavor and taste, consistency) [22, 27–30].

The starch gelatinization index determines the intensity of heat treatment in a given material. The extruded food products do not display 100% gelatinized starch because part reacts with other food ingredients and creates insoluble, unavailable complexes, which can be regarded as ballast for the organism [18].

The degree of starch gelatinization in extruded precooked products may reach a level of 80–98% if the process parameters, including the water content and fine

Figure 7.2 A modified single-screw food extruder [23].

grinding of the ingredients, are properly adjusted. The result of the process may be instant starches soluble in cold water or products not requiring cooking before consumption [4, 12, 13, 15, 26]. Dry and hydrated precooked pasta are shown on the Figure 7.3.

In conventional pasta pressed at low temperatures (below 50 °C) the degree of starch gelatinization after drying does not exceed 50%; therefore, traditional pasta is cooked before serving. Pasta designed to have a short thermal treatment by steaming, pre-cooking, toasting or hydration displays a higher degree of gelatinized starch (Figure 7.4). A gelatinization level of 95% can be obtained in extruded pasta after applying additional cooking in a hot water bath and drying. In instant pasta products subjected to steaming and frying, the quantity of gelatinized starch was estimated at 84–88%, depending on the frying parameters [2].

In extrudates produced from different types of flour at a temperature below 100 °C the observed starch gelatinization index was 82–88%. At the same time, the increase in water content also raised the level of gelatinized starch [7, 14]. The starch gelatinization increased with the intensity of the thermo-mechanical treatment inside the extruder operated with the SME input ranging from 0.02 to 0.58 kWh per kg of the product [10, 15].

The application of functional additives also alters the characteristics of starch gelatinization in pasta products obtained from different flour types. It was observed that after the addition of methylcellulose the amount of gelatinized starch in the

Figure 7.3 Extrusion-cooked wheat pasta: dry and ready for consumption [14] .

products increased from 79% (with the addition of 0.02% methylcellulose) to nearly 90% (with the addition of 0.10%). In contrast, the opposite effects were observed when determining the degree of starch gelatinization in products enriched with L-ascorbic acid [1, 16]. The increase in acid content in the material mixture reduced gelatinization, which may be associated with the functional nature of L-ascorbic acid strengthening the structure of gluten, and thus consolidating the carbohydrate and protein complexes emerging in thermal processes [23]. Application of low screw rpm during pasta processing at a temperature below 100 °C using a single-screw extruder resulted in a low level of gelatinized starch, not exceeding 80%, so the final product was unstable during hot water hydration and its texture was poor [7, 15].

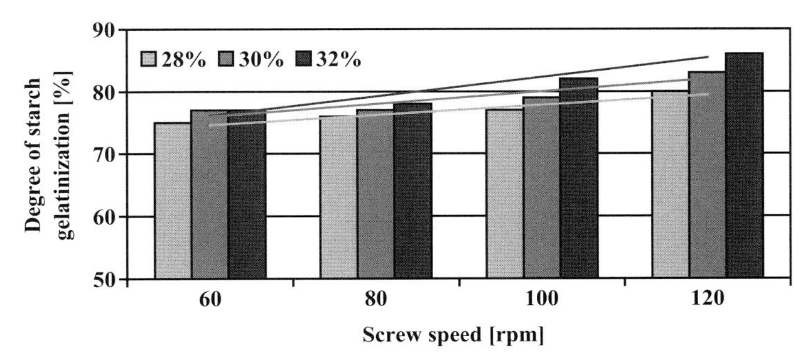

Figure 7.4 Starch gelatinization degree in precooked pasta made from semolina with different moisture levels [7].

Both the dough moisture and extrusion conditions were found to affect the water absorption of pasta made from common wheat flour. Water uptake is an indicator of the thermal treatment intensity and the ability of starch to absorb water. The most suitable products were obtained using pasta dough with a moisture content of 30%, while the water absorption was reduced with higher initial dough moisture and the application of higher screw rpm during extrusion [1, 16]. This parameter is also influenced by functional additives. Pasta with monoglyceride addition showed water absorption ranging from 200 to 270%, decreasing with a higher amount of emulsifier in the recipe (maximum 1%) [21]. Pasta processed with the addition of wheat bran showed higher water absorption (from 230 to 290%) on increasing the amount of bran in the recipe from 5 to 25%, respectively [26]. The addition of legumes in the range 10–40% increased the water absorption from 250 to 340% due to a higher protein content in the processed pasta products (from 11.9 to 17.8%, respectively) [11]. Starchy precooked pasta processed with rice or corn flour had water absorption in the range 280–340% for corm pasta and 300–400% for rice pasta [12]. Water absorption may suggest the amount of water that should be added to precooked pasta during hot water hydration to give the proper consistency.

Instant or precooked pasta products should exhibit a low level of expansion, since this is related to the number of pores and cracks on the surface of the product and to reduction in the hydration time. The radial expansion index was entirely dependent on the extruder's work parameters; regardless of the raw materials used, the higher the screw speed applied, the higher the rate of expansion. This parameter was also dependent on functional additives. The application of methylcellulose reduced the value of the radial expansion with increasing percentage of the additive [16]. On the other hand, the addition of L-ascorbic acid had the reverse effect [23]. The increasing addition of legumes led to a reduction in the expansion [10, 11]. The same tendency was observed when bran fibers were added to the recipe [26]. Gluten-free corn and rice pasta showed a lower expansion ratio (1.1. to 1.6) when using an $L/D = 18$ barrel configuration with an intensive cooling system applied to the final barrel section [12].

Compared with traditional pasta whose cooking time is 8–10, or even 20 min, the precooked extruded products exhibited shorter preparation times with no cooking involved and the hydration time in hot water ranged from 4 to 8 min [11, 12, 15, 16, 21, 26].

The cooking losses are an important parameter in determining the functional pasta characteristics. The smaller the amounts of ingredients leaching into the water after cooking, the better the quality of the pasta. For instant or precooked products that do not require cooking and where the preparation time is very short, the quantity of ingredients rinsed out from the pasta or noodles containing starchy raw materials should not exceed 10% [2, 3, 8, 9, 19, 25, 29].

The main factors influencing the cooking losses after hydration were the applied screw rpm and the degree of starch gelatinization (Figure 7.5). With the application of intensive mechanical treatment, an appropriate moisture level of semolina and a high amount of gelatinized starch, a lower level of cooking loss was noted [15]. In precooked pasta enriched with legumes (pea, bean and lentil addition from 10 to 40%) cooking losses after the 5 min hydration were less than 9%, and increased with

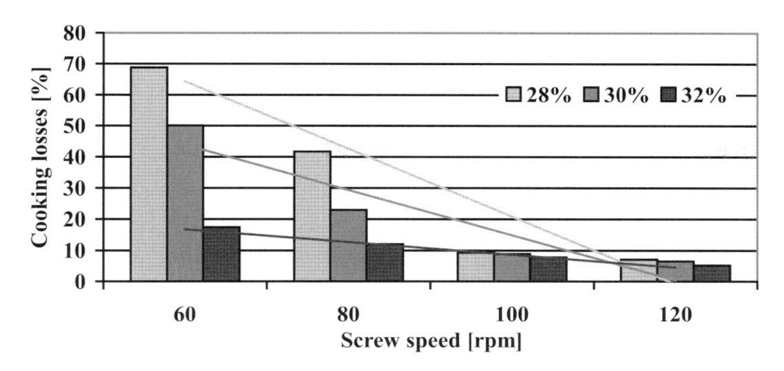

Figure 7.5 The cooking losses during hydration of precooked pasta made from semolina with different screw rpm used [7].

higher concentrations of pulses in the recipe [10, 11]. Increased addition of wheat bran led to the highest cooking losses, even reaching 15% with the addition of 25% bran [26]. Cooking losses of gluten-free precooked pasta varied from 4 to 8% and were strongly dependent on the screw speed during the extrusion-cooking for both corn and rice products. The moisture content of the raw materials showed low influence on this parameter, but low moisture combined with low extrusion speed led to the highest amount of compounds present in the water (14%) [12].

The results presented by many researchers indicate the importance of this parameter for the characteristic of the pasta. Depending on the flour type and thermal treatment of the dough, the cooking losses varied between 3.7 and 11.6% [17]. At various protein content in flour, Edwards *et al.* [27] reported cooking losses from 7.1 to 15.5%, while Debbuoz and Doekott [25], depending on the extrusion parameters, found losses from 5.6 to 6.1%. The most considerable differences in measurements were seen by Abecassis *et al.* [24] who estimated the cooking losses from 9.6 to 51.5% in pasta produced with different process parameters. They also argued that the culinary usefulness of the pasta was determined by the combined impact of extrusion temperature, material moisture and speed of extrusion. Greater losses of ingredients during cooking may also be the result of damage to the starch granules from mechanical treatment and, consequently, slower water penetration, stickiness of the pasta surface, low quality of the flour and darker color of the products [3].

The texture of pasta products plays a main role in the consumers' assessment of its quality and attractiveness. Both the appearance of dry products offered in trade, as well as the quality determinants during and after cooking, are important parameters that affect the products' overall evaluation. Some other pasta quality features are firmness, stickiness, elasticity, adhesiveness, chewiness and bulkiness whose interactions depend both on the material composition and the production technology [22, 27, 28, 30, 31].

Matsuo and Irvine [28] used materials with different protein composition for the manufacture of pasta whose texture was subject to testing by an Instron device equipped with a unique head (Plexiglas tooth) that simulates the bite. This method

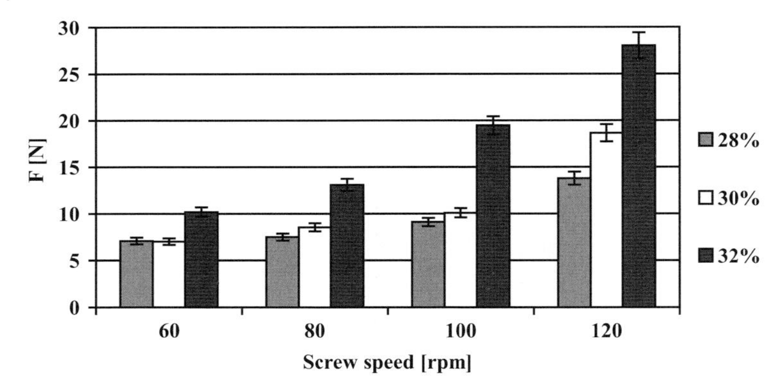

Figure 7.6 The hardness of pasta made from semolina at different initial humidity content and different rotational speed of screw [14].

lets to the definition of firmness as the work needed to complete mastication of a spaghetti strand. There are many results based on this method of determining the pasta texture characteristic [3, 5, 9, 27, 28, 30]. Similarly, the use of the Kramer cell in testing machines allows to evaluate the hardness and firmness of pasta as the breaking strength and the work needed to break the pasta structure and also other textural parameters like chewiness, springiness or adhesiveness of food products [2, 28, 30].

The cutting force can be interpreted as the hardness of pasta products in the texture evaluation (Figure 7.6). The texture of the pasta depends upon the dough moisture content and the conditions of thermal and pressure processing related to the physiochemical changes [22, 25, 27, 28, 30, 31]. Depending on the region of the world and the type of pasta, consumers most appreciate the so-called *al dente* slightly hard pasta or soft noodles with low hardness [3, 18, 27].

Figure 7.7 shows one of the diagrams illustrating both the cutting forces and the deformation after subsequent minutes of precooked pasta hydration. It was observed that for all of the products the largest cutting forces and smallest deformation values occurred after the first minute of hydration, and this relationship changed with further time of pasta hydration. After few minutes of hydration the cutting force was the lowest and deformation the greatest, irrespective of the additives used [14, 23, 26].

Sensory characteristics are equally important in the evaluation of the quality of food products. Organoleptic features differed for precooked products depending on the raw materials and additives used and the process conditions applied. The appearance, color, flavor and consistency were noted in a five-point scale. Pasta processed with low screw rpm exhibited poor transparency, glassiness, more floury fracture and a desirable in terms of hydration feature – thin threads – making them brittle and susceptible to fracture. It lowered the organoleptic evaluation rating to 2 points for the color and appearance and 4 points for the flavor [1, 16]. It was observed that, depending on the raw materials and additives used, the products manufactured at 30% dough moisture content and medium screw rpm during the extrusion process at the proposed range of temperatures showed the highest

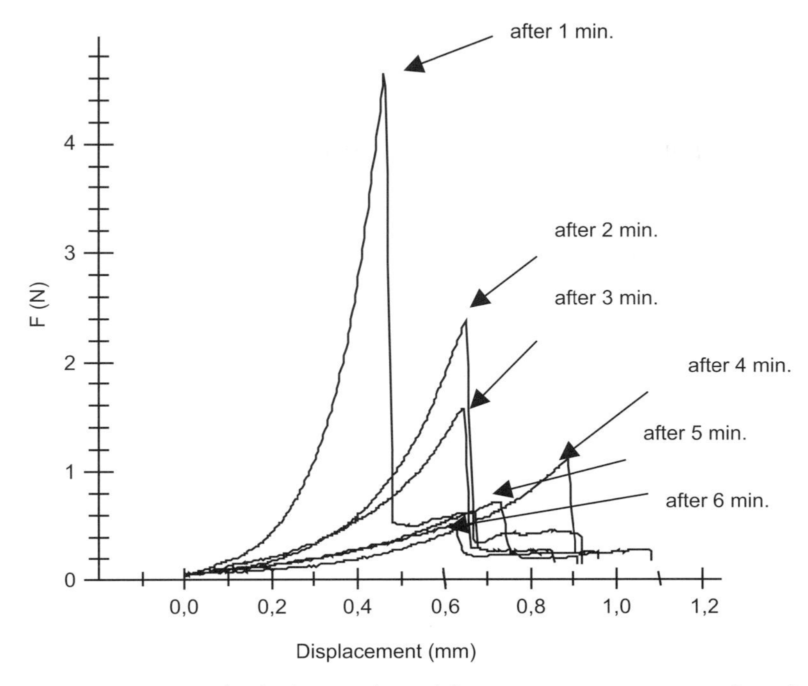

Figure 7.7 An example of a diagram obtained during a texture measurement of semolina precooked pasta, processed at 120 rpm with 28% dough moisture content hydrated for different times [14].

desirability at the level of positive and high acceptance. The consistency of all the products processed at these parameters was firm, their surface smooth, color even and taste sensations close to conventional pasta products [1]. An example of sensory evaluation of gluten-free pasta processed at different conditions is presented in Table 7.2.

The influence of an additive, that is, wheat bran, had an impact on the sensory scores: a higher amount of bran led to lower sensory scores associated with the presence of harder in bite particles of bran and a specific taste related to the floury taste of wheat bran. Application of higher rpm during extrusion-cooking of pasta-like products improved the sensory quality, restricted stickiness and improved the shape keeping and taste of the products. This may be explained by better association of components under intensive extrusion-cooking treatment (higher shearing stress), which is able to form the starch–protein matrix during processing at a temperature higher than the starch gelatinization temperature and in the presence of 30% moisture content in the raw materials. The overall quality of pasta-like products enriched with wheat bran was lowered, especially when the bran level in the recipe was higher than 20% (Table 7.3). Darker color was observed for samples enriched with 20 and 25% of wheat bran, the shiny and smooth surface of the pasta strands disappeared; products became sticky and lost their consistency.

Table 7.2 Sensory evaluation of precooked gluten-free pasta processed at different parameters [12].

Moisture content [%]	Screw speed [rpm]	Corn pasta			Rice pasta		
		Appearance	Taste	Stickiness	Appearance	Taste	Stickiness
30	60	3.0	3.0	3.0	2.5	1.5	1.5
	80	4.0	4.0	4.0	3.0	3.0	3.0
	100	5.0	5.0	5.0	4.0	4.0	4.0
	120	5.0	5.0	5.0	4.0	5.0	5.0
32	60	4.0	4.0	3.5	3.0	3.0	2.0
	80	4.0	4.5	4.0	3.5	3.5	3.0
	100	4.5	4.5	5.0	4.0	4.0	4.0
	120	5.0	5.0	5.0	5.0	4.5	5.0
34	60	3.0	3.0	3.0	3.0	3.0	2.0
	80	3.0	3.0	3.0	3.0	3.0	2.0
	100	3.5	3.5	3.5	3.5	3.5	2.5
	120	4.0	4.0	4.0	4.0	4.0	4.0

A very important factor for understanding the changes during the extrusion-cooking process of precooked pasta is the evaluation of the microstructure by means of scanning electron microscopy (SEM). The evaluation was based on the microscopic analysis of the dry products carried out at different magnifications (Figure 7.8), with similar analysis of the products after a 4 min hot water hydration (Figure 7.9). They allowed analysis of the microstructure of products processed using different recipes and different extrusion parameters [16].

Pasta products made from wheat flour type 500 with 28% dough moisture content and screw rotation of 60 rpm exhibited a disorganized structure without clearly marked starch fractions. In pasta made from flour with a moisture content of 30% the development of a matrix linking the granules of swelled starch was observed. A dough with 32% initial moisture content extruded at 120 rpm showed a gelatinized starch with compact and homogenous structure visible on the image. Another internal structure was observed in pasta products made from semolina using various screw

Table 7.3 Overall sensory assessment of pasta enriched with wheat bran processed at different screw speeds with initial moisture content 30%.

Screw speed [rpm]	Wheat bran addition					
	Without additive	5%	10%	15%	20%	25%
60	4.3	3.8	3.6	3.4	2.8	2.2
80	4.7	4.6	4.2	3.9	3.1	2.5
100	4.9	4.9	4.7	4.2	3.2	2.7
120	4.9	4.8	4.8	4.4	3.8	3.4

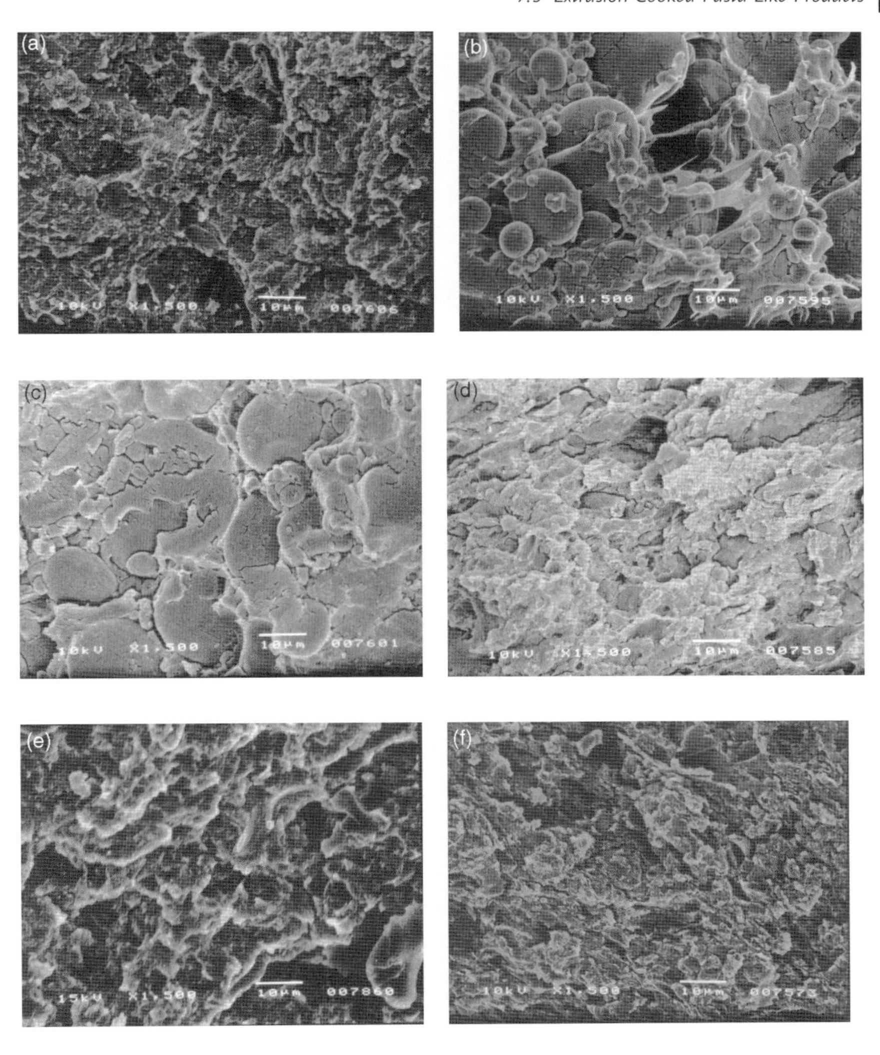

Figure 7.8 A cross-section of dry precooked pasta made from common wheat flour type 500 with different processing parameters (magnification x1500) [1]: (a) 28% m.c., 60 rpm, (b) 30% m.c., 80 rpm, (c) 32% m.c., 100 rpm, (d) 0.1% methylocellulose, 30% m.c., 80 rpm, (e) 0.1% ascorbic acid, 30% m.c., 80 rpm, (f) semolina, 30% m.c., 120 rpm.

rotations during the extrusion. Pasta made at the slowest rpm had a broken, disrupted structure with separate starch granules; on the other hand, pasta processed at a high screw speed exhibited an ordered and coherent structure, free from slots and separated starch granules [15].

The application of functional additives or raw materials other than wheat also had an impact on the physiochemical characteristics of the pasta, and thus on the internal pasta structure. In pasta extruded from flour type 500 with the addition of

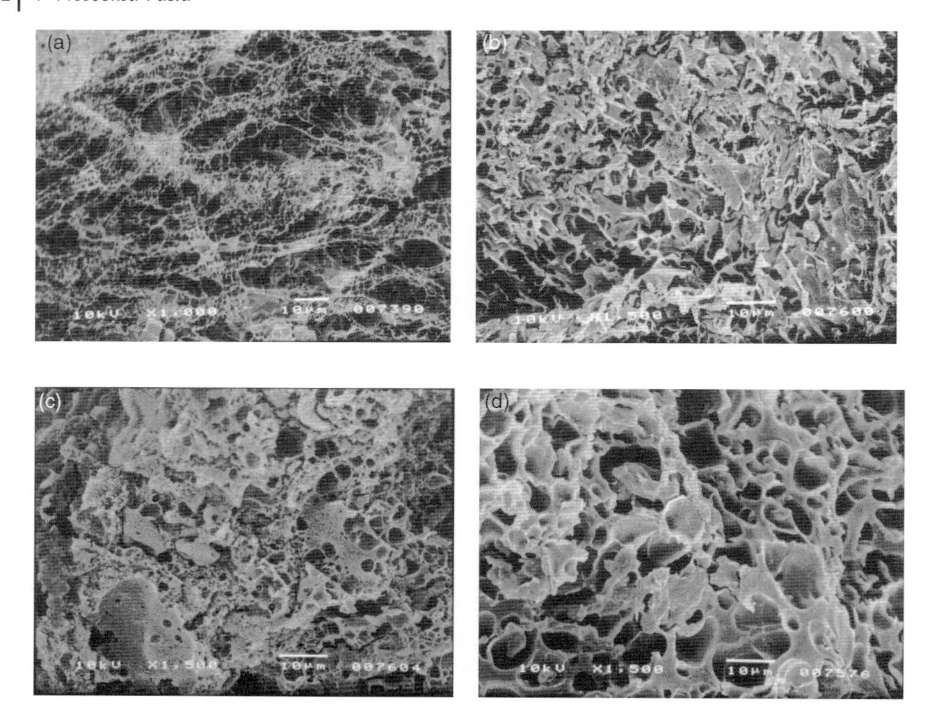

Figure 7.9 Microstructure of selected pasta products after hydration (magnification x1500) [16]: (a) surface of semolina pasta, 30% m.c., 80 rpm, (b) surface of common wheat flour pasta, 28% m.c., 60 rpm, (c) cross-section of common wheat flour pasta, 32% m.c., 100 rpm, (d) cross-section of semolina pasta, 30% m.c., 120 rpm.

methylcellulose some changes were observed in the internal structure of the pasta compared with the pasta without the addition of this component. The gluten consolidated with swelled starch granules clearly disappeared and, with a small addition of methylcellulose, the internal structure was more compact with a small amount of free space. A homogeneous character of the internal structure was observed in pasta extruded at 80 rpm. A corrugated and porous structure was observed inside products made from common wheat flour with the addition 0.1% of L-ascorbic acid (Figure 7.8.e) [1, 16].

Similar images with slightly more explicit starch granules were observed by Cunin *et al.* [31]. They analyzed semolina pasta produced on a single-screw pasta press at a temperature of 40 °C, equipped with a Teflon die with openings 1.9 mm in diameter. Thorvaldsson *et al.* [20], when testing the transition of pasta dough from wheat flour with a moisture content of 39.5 and 41% processed at increasing treatment temperature from 20 to 85 °C, showed a clear increase in dough density with increase in temperature and humidity; the entire disappearance of the initially visible starch granules was observed at the highest temperature, due to total starch gelatinization under these conditions.

The SEM pictures of the internal precooked pasta structure after hot water hydration showed differences on the surface of singular pasta threads processed from different recipes and at various extrusion-cooking parameters. Microscopic images with magnification x1500 (Figure 7.9) showed intensive water penetration to the pasta center during hydration and the formation of a porous surface structure with visible cavities occurred during leaching of unbound components from the pasta – the dark places indicate cavities left by the lost ingredients after hydration.

The penetration of hot water into the pasta leads to the entire gelatinization of unbound starch and the formation of a honeycomb-like protein–starch structure, typical for extruded or expanded products [1, 31]. Pasta products obtained from mixtures with low moisture content extruded at 60 rpm showed fine, regular pores, although the cooking losses during hydration were up to 10% of the pasta weight. A different structure was observed for hydrated semolina pasta. Their surface revealed large, deep pores which, in some areas, began to convert into a gluten grid, not yet preventing the loss of components. The losses during hydration were considerable, reaching up to 50%, when a low screw speed was applied [15].

The microstructure of dry precooked corn pasta showed a dense and compact internal structure with only few disruptions due to improper mixing or the presence of steam inside the pasta threads (Figure 7.10). In pasta processed at low moisture content singular untransformed starch granules were visible, increasing the screw speed limited this tendency. An application of higher initial moisture eliminated this effect and the internal structure was regular [12].

A similar internal structure was observed for rice precooked pasta, a dense and compact structure was observed for almost all tested samples. Also, in these products applications of low moisture and low extrusion speed generated the presence of small areas with visible untreated starch granules. A higher screw speed leads to the formation of some aggregates of swollen but not completely gelatinized starch. The internal structure of pasta threads processed at higher rpm and using raw materials with higher moisture content was compact and regular (Figure 7.11).

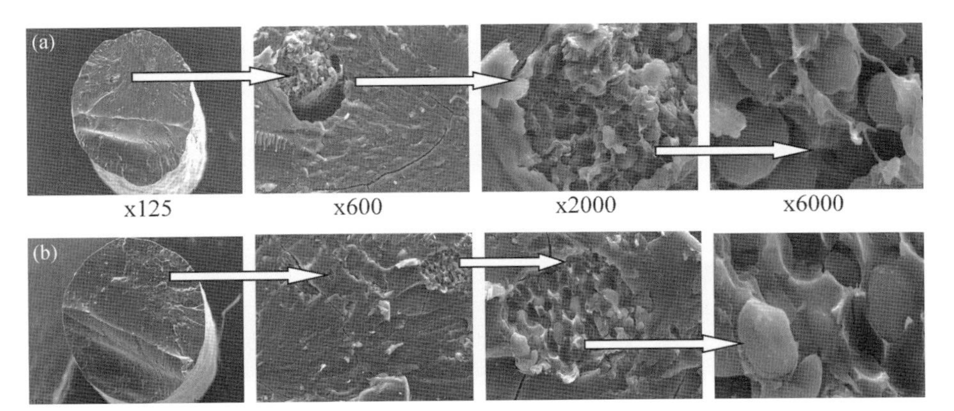

Figure 7.10 Microstructure of corn pasta processed at 30% initial moisture content at different screw speeds: (a) 60 rpm, (b) 120 rpm, at different magnifications.

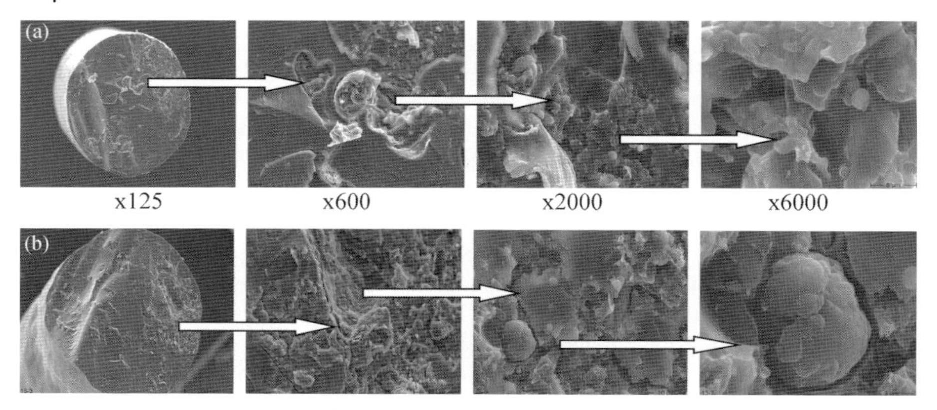

Figure 7.11 Microstructure of rice pasta processed at 30% initial moisture content at different screw speed: (a) 60 rpm, (b) 120 rpm, at different magnifications.

The internal microstructure of precooked pasta enriched with wheat bran showed a compact and dense structure formed as a starch and protein matrix with only few ungelatinized starch granules visible close to the bran parts (Figure 7.12a). Low screw speed during processing led to a more varied internal composition, observed as cross-sections of pasta, than for highest rpm used. So the impact of screw speed and thus treatment intensity on the internal structure is significant. Observations of precooked pasta enriched with 5% wheat bran showed a compact internal structure with rarely visible singular wheat bran unconverted parts.

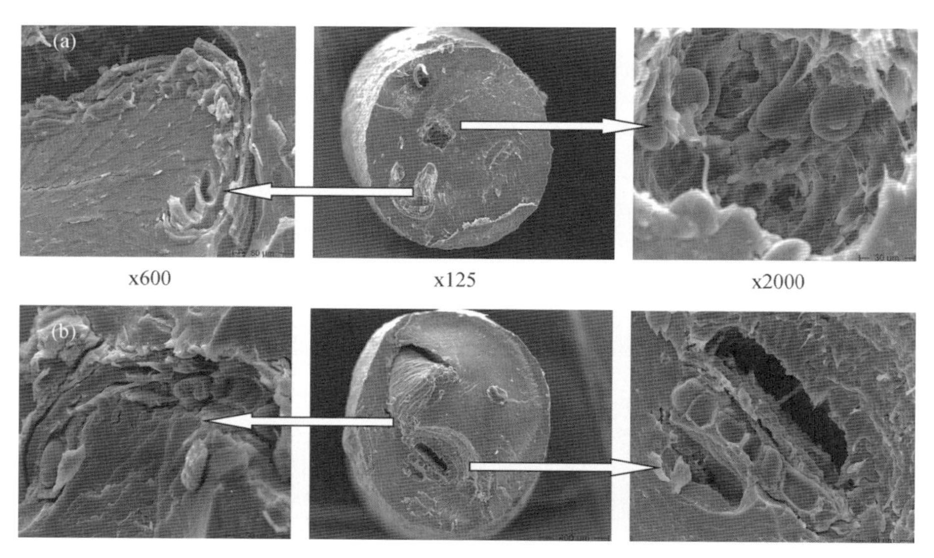

Figure 7.12 Microstructure of wheat bran enriched precooked pasta processed at 30% initial moisture content at different screw speed: (a) 5% wheat bran, 60 rpm, (b) 20% wheat bran, 100 rpm, at different magnifications.

SEM pictures of precooked pasta enriched with 20% of wheat bran showed higher amount of bran particles in the inside structure, it was clearly visible untreated bran surface and cellular character of outside fiber layer (Figure 7.12b). Wheat bran composition is fibrous and during the thermo-mechanical treatment by extrusion-cooking it was observed separations of singular fibers from bran structure. Also loose and empty structures placed close to fiber parts in the inside was observed, so it may be the reason of weaker structure and lower hardness of pasta enriched with high addition of wheat bran [26].

7.4 Conclusions

Summarising the results presented above, the use of a modified single-screw extruder type TS-45 Polish design equipped with an extended plasticization unit (L/D = 16 or 18) and an additional cooling system offers the possibility to process good quality precooked pasta products from common wheat flour with functional additives or from starchy raw materials like rice or corn flour. The temperature range of thermal treatment of wheat flour within 70–100 °C and the screw speed of 80–120 rpm during the processing allowed the manufacture of products with a short preparation time and ready to eat after 3–5 min of hot water hydration. The extruded products showed a high starch gelatinization degree compared with traditional products, so that they can be considered as precooked foodstuff. The microstructure of the pasta observed with SEM confirmed different internal structures depending on the raw materials used and the production parameters. A homogeneous and compact structure was observed in pasta processed at a higher screw rpm. The consistency and other sensory characteristics of products processed at 80–100 rpm were correct for precooked pasta, the texture was firm, the uniform surface, and taste sensations on consumption were similar to those experienced with traditional pasta. So application of the extrusion-cooking technique may be successful for processing a wide range of new types of fat-free, enriched precooked pasta.

References

1 Wójtowicz, A. (2003) Evaluation of extrusion-cooking processing of precooked pasta. PhD Thesis, Akademia Rolnicza, Lublin (in Polish).

2 Kruger, J.E., Matsuo, R., and Dick, J. (1996) *Pasta and Noodle Technology*, American Association of Cereal Chemistry Inc., USA.

3 Li, J. and Vasanthan, T. (2003) Hypochlorite oxidation of field pea starch and its suitability for noodle making using an extrusion cooker. *Food Res. Inter.*, **36** (4), 381–386.

4 Oh, N. and Sheen, S. (2002) Instant pasta with improved cooking quality, Patent WO/2002/045526.

5 Wang, N., Bhirud, P., Sosulski, F., and Tyler, R. (1999) Pasta-like product from pea flour by twin-screw extrusion. *J. Food Sci.*, **4**, 671–678.

6 Oh, N., Seib, P., Ward, A., and Deyoe, C. (1985) Noodles IV. Influence of flour

protein, extraction rate, particle size, and starch damage on quality characteristics of dry noodles. *Cereal Chem.*, **62** (6), 441–446

7 Wójtowicz, A. (2006) Influence of semolina moisture content and process parameters on some quality parameters of precooked pasta. *Acta Agrophys.*, **8** (1), 263–273 (in Polish, abstract in English).

8 Kim, Y., Wiesenborn, D., Lorenzen, J., and Berglund, P. (1996) Suitability of edible bean and potato starches for starch noodles. *Cereal Chem.*, **73** (3), 302–307.

9 Manthey, F., Yalla, S., Dick, T., and Badaruddin, M. (2004) Extrusion properties and cooking quality of spaghetti containing buckwheat bran flour. *Cereal Chem.*, **81** (2), 232–236.

10 Wójtowicz, A. (2008) Influence of legumes addition on proceeding of extrusion-cooking process of precooked pasta. TEKA Commission of Motorization and Power Industry in Agriculture, 8a, 209–216.

11 Wójtowicz, A. (2009) Influence of white bean addition on selected parameters of extruded precooked pasta. *Acta Agrophys.*, **13** (2), 543–553 (in Polish, abstract in English).

12 Wójtowicz, A. and Mościcki, L. (2009) Evaluation of selected quality characteristics and texture of gluten-free precooked pasta. Abstract Book, 5th International Congress Flour-Bread'09, Opatija, Croatia, p. 34.

13 Huber, G.R. (1988) Extrusion cooking applications for precooked pasta production, Wenger Manufacturing Co.

14 Wójtowicz, A. (2006) Influence of extrusion parameters on some texture characteristics of precooked semolina pasta. *Acta Agrophys.*, **8** (4), 1049–1060 (in Polish, abstract in English).

15 Wójtowicz, A. and Mościcki, L. (2009) Influence of extrusion-cooking parameters on some quality aspects of precooked pasta-like products. *J. Food Sci.*, **74** (5), E226–E233.

16 Wójtowicz, A. (2005) Influence of some functional components addition on microstructure of precooked pasta. *Pol. J. Food Nutr. Sci.*, **14/55** (4), 417–422.

17 Fardet, A., Abecassis, J., Hoebler, C., Baldwin, P., Buleon, A., Berot, S., and Barry, J. (1999) Influence of technological modifications of the protein network from pasta on in vitro starch degradation. *J. Cereal Sci.*, **30**, 133–145.

18 Pagani, M.A. (1986) *Pasta and Extrusion Cooked Foods*, Elsevier Applied Science Publishers, London and New York.

19 Malcolmson, L. and Matsuo, R. (1993) Effect of cooking water composition on stickiness and cooking loss of spaghetti. *Cereal Chem.*, **70** (3), 272–275.

20 Thorvaldsson, K., Stading, M., Nilsson, K., Kidman, S., and Langton, M. (1999) Rheology and structure of heat-treated pasta dough: influence of water content and heating rate. *Lebensm. -Wiss. U Technol.*, **32** (3), 154–161.

21 Wójtowicz, A. (2007) The influence of monoglyceride and lecithin addition on some cooking quality of precooked pasta. *Pol. J. Food Nutr. Sci.*, **14/55** (4), 417–422.

22 Rho, K., Seib, P., Chung, O., and Deyoe, C. (1988) Noodles. VII. Investigating the surface firmness of cooked oriental dry noodles made from hard wheat flours. *Cereal Chem.*, **65** (4), 320–326.

23 Wójtowicz, A. (2004) Influence of ascorbic acid on texture of extruded precooked pasta. *Acta Agrophys.*, **4** (2), 589–599 (in Polish, abstract in English).

24 Abecassis, J., Abbou, R., Chaurand, M., Morel, M.H., and Vernoux, P. (1994) Influence of extrusion conditions on extrusion speed, temperature, and pressure in the extruder and on pasta quality. *Cereal Chem.*, **71**, 247–253.

25 Debbouz, A. and Doetkott, C. (1996) Effect of process variables on spaghetti quality. *Cereal Chem.*, **73** (6), 672–676.

26 Wójtowicz, A. (2009) Influence of wheat bran addition on functional and sensory characteristics of precooked pasta products. Book of Abstracts, 17th Internationale Conference of IGV GmbH, Bergholz-Rehbrücke, Germany, p. 39.

27 Edwards, N., Izydorczyk, M., Dexter, J.E., and Biliaderis, C. (1993) Cooked pasta texture: comparison of dynamic viscoelastic properties to instrumental assessment of firmness. *Cereal Chem.*, **70** (2), 122–126.

28 Matsuo, R. and Irvine, G. (1971) Note on an improved apparatus for testing spaghetti tenderness. *Cereal Chem.*, **48**, 554–558.

29 Matsuo, R., Malcolmson, L., Edwards, N., and Dexter, J.E. (1992) A colorimetric method for estimating spaghetti cooking losses. *Cereal Chem.*, **69** (1), 27–29.

30 Smewing, J. (1997) Analyzing the texture of pasta for quality control. *Cereal Foods World*, 42 (1), 8–12.

31 Cunin, C., Handschin, S., Walther, P., and Escher, F. (1995) Structural changes of starch during cooking of durum wheat pasta. *Lebensm. -Wiss. U Technol.*, **28**, 323–328.

8
Processing of Full Fat Soybeans and Textured Vegetable Proteins

Leszek Mościcki

8.1
Introduction

Research on the potential application of a thermal treatment in raising the utility value of vegetable materials has shown numerous advantages of extrusion-cooking and expander-cooking technology in the processing of the seeds of legumes and oil plants for the purposes of food and feed manufacture. For example, a baro-thermal treatment of full-fat soybeans or raw rapeseed enables the production of a protein-rich, highly nutritious feed component whose benefits certainly off set the costs incurred for enrichment of the above-mentioned raw materials [1–3]. Its application offers the following:

- inactivation of the antinutritional factors contained in soybeans and legumes,
- improvement of product taste,
- denaturation of protein and reduction of its degree of solubility,
- increase in availability and use of fat and inactivation of lipoxidase.

The size of soy cells is between 30 and 50 microns. These cells contain proteins with a diameter of 6 to10 microns and lipids of diameter 0.2 to 0.5 microns. During an extrusion-cooking of soybean, lipids are combined into larger drops of fat, protein chains split and denaturation of the proteins follows. Cell walls also split and the contents of various cells are released and merge into one uniform mass.

Knowledge of the cellular structure of soybean and its careful processing can be used to obtain a complex final product. Soy extrudate, used as a feed component in mixtures for monogastric animals, must exhibit other functional properties than the extrudate prepared to be used as a milk replacer for calves [3]. The extruded full-fat soy intended as an additive to fish feed requires different conditions of thermal and pressure treatment compared with that to be used in the manufacture of feed for broilers or laying hens. It is the possibility of altering the process parameters that enables the manufacture of such a wide range of products.

Research has shown that maximum efficiency and productivity of a process is achieved when soy seeds are fragmented into particles of diameter less than

Extrusion-Cooking Techniques: Applications, Theory and Sustainability. Edited by Leszek Moscicki

ISBN: 978-3-527-32888-8

2 mm [1, 4]. It was further concluded that, in order to obtain high-quality meal, seeds should be ground no more than 2 hours before extrusion. The results of a relatively rapid processing of soybeans are minimum liberation of free fatty acids and prevention of negative phenomena such as fat rancidity.

8.2 Extrusion-Cooking of Full-Fat Soybeans and Other Protein-Rich Vegetable Materials

As already mentioned, the thermal treatment of full-fat soybeans depends to a large extent on the final purpose of the received product. For this reason, three main treatment methods can be distinguished for obtaining appropriate enrichment effects.

The first method consists in implementing the most moderate thermal treatment. Growth inhibitors can be inactivated by supplying an adequate quantity of steam to the raw material and using less intensive mixing. The aim of this process is to neutralize the activity of trypsin to less than 1.0 trypsin inhibitory units (TIU)/mg (nutritionists regard the value of trypsin at 5.0 TIU/mg as safe because it corresponds to the destruction of 90% of growth inhibitors) [24].

The requirements of the second method concern not only the destruction of growth inhibitors but also the simultaneous release of oil contained in the seeds. This process needs the delivery of a greater amount of energy, especially mechanical energy. During a deeper baro-thermal treatment, lipid–protein complexes and hydrocarbon–lipid complexes are developed which are available for monogastric animals.

The third and most intense method is, in this case, extrusion-cooking, used specifically for the protection of fatty complexes (tolerated by ruminants) and for denaturation of protein. Due to the extended duration of thermal treatment, the processed soybean involves the denaturation of the maximum quantity of protein; this may lead to the phenomenon of product browning. It can be controlled by tracking the NSI index (nitrogen solubility index) which helps determine the degree of solubility of protein that may go directly into the abomasums of ruminants during digestion.

Nowadays, two extrusion models are available on the market: dry extrusion and moist extrusion, each with its own advantages and disadvantages. Dry extrusion (based on autogenic extruders) subjects the ground beans to pressure inside a barrel powered by rotating a single screw. The pressure reaches a level of 3.5–4.0 MPa and the heat produced by the friction between the beans and the walls of the barrel heats and sterilizes the product. This treatment takes less than 20 s and reaches temperatures of 120 to 165 °C, depending on the device used [5, 6]. The main disadvantage of this treatment method is that the friction can cause excessive temperatures which in turn influence the level of available lysine. Dry extrusion is still more popular than moist extrusion, especially in the USA, as it is less expensive and farmers can use it on-site to treat their own soybeans [6]. The capacity of this process varies from a few hundred kilos to several tons per hour, depending on the size and whether or not any preconditioning is carried out [7].

A typical moist extrusion plant comprises a seed cleaner, a preconditioner, a feeder, a barrel complete with bolts and steam injection valves and one or twin screws of similar configuration powered by an electric motor, a drier and a cooler. It is advisable to use a bean mill with a very fine mesh (if possible less than 1.0 mm) in order to guarantee the homogeneity of the particles. Within the preconditioning unit, steam is added to take the mixture to a moisture level of 24 to 28% and a temperature of 80 to 90 °C. The size of the screw depends on the distance concerned so as to obtain an optimally homogeneous mixture and to ensure that the pressure applied to the mixture is sufficiently high. The pressure level inside the barrel is around 3.0 MPa such that the water does not evaporate in spite of the high temperatures it can reach. When it leaves the extruder, the mixture "blows", the water rapidly evaporates and, as a consequence, the oil cells burst, releasing the oil. However, this oil is absorbed once again as the mixture is cooled and remains locked inside. Next, the bean is placed inside the drier for 14 min where its moisture level is reduced to 14 to 16% before it is placed inside a horizontal cooler, reducing its final moisture level to between 10 and 12%.

Moist extrusion installations are more expensive than those used for dry extrusion. On the other hand, they have a superior production capacity and are more effective in terms of denaturing the antinutritional factors [1, 3].

Perilla *et al.* [5] studied the influence of the temperature used for moist extrusion on the antinutritional factors of the beans and the productivity of chickens. The machine used was an Anderson extruder and the average residence time chosen was 20 s. The authors indicated that the optimum temperature for this machine is between 122 and 126 °C, very close to that recommended by the manufacturer. Clarke and Wiseman [8] using a twin-screw extruder studied the influence of the temperature and moisture levels present during the extrusion process on the quality of the beans and its feeding effect in growing chickens. The antitrypsin activity fell as the moisture and temperature were increased. Where no additional moisture was added, the residual antitrypsin activity was greater than for those beans to which 14 to 24% extra moisture was added. Indeed, the antitrypsin activity of those beans extruded at 70 °C without additional moisture was similar to that of raw beans (35 vs. 37.5 mg/g of beans), while temperatures greater than 115 °C produced acceptable values for all of the samples studied (<5 mg/g). The authors concluded that temperatures of 130 to 160 °C combined with moisture levels of 11 to 35% are sufficient to obtain quality beans. The antitrypsin activity values obtained during this study and expressed in mg/g where 15% extra water was added to the beans were 23 at 70 °C, 15 at 90 °C, 6 at 110 °C and 3 mg/g at 130 and 150 °C. The best productivity results for chickens were obtained with beans processed at the highest temperature.

Zarkadas and Wiseman [9] extruded beans at temperatures of between 70 and 150 °C which they then fed to piglets of 10 to 27 kg. They noted that the best productivity results were obtained with soybean trypsin inhibitor levels lower than 5 mg/g, which corresponds to ingestion of less than 1.5 mg/d. These values were only obtained with extruder temperatures of greater than 110 °C.

A variety of extruders designed for soy processing are available on the market. Single screw extruders are still in the majority, however many feed producers have moved to versatile twin-screw units. The current trend is towards increasing production levels per installation and reducing the related energy costs. This requires better preconditioning of the initial product, a more homogeneous mixture and a better L/D ratio. The moist extrusion process is more effective, offers more optimum control, has lower costs in terms of energy and replacement parts and provides a more homogeneous end product than dry extrusion, despite the fact that the end product requires to be dried.

Extrusion-cooking of full-fat soybeans can be carried out on a processing line, such as presented in Figure 8.1, based on a single-screw extruder of L/D = 25, equipped with a large conditioner. In order to obtain a product of good quality, the extrusion-cooking process should be held at a temperature of 120–135 °C and a moisture content of 17–20%. It was proved that the degree of deactivation of growth inhibitors in the processed soybean depends on the temperature and time of the baro-thermal treatment. Preconditioning before extrusion-cooking considerably helps deactivation of growth inhibitors and significantly increases the efficiency of the process. Maintaining a higher temperature of the product immediately after extrusion also contributes to the effectiveness of thermal treatment. Thus, it is advisable to transport a hot extrudate directly from the die to the cooler by, for example, an encased, slow-speed screw conveyor, which allows the product to be held at a higher temperature for some additional time.

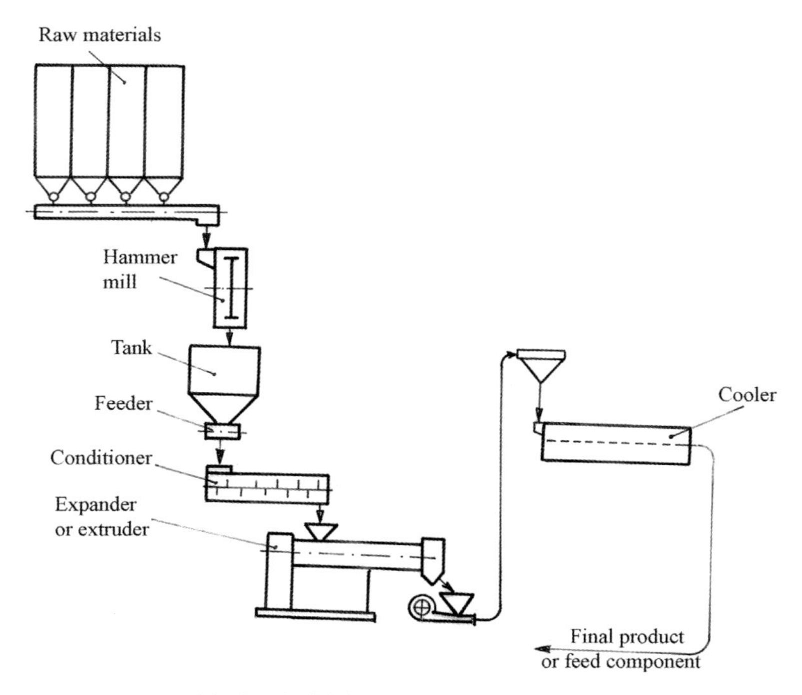

Figure 8.1 Set-up of the line for full fat soybeans processing.

As for the quality of the finished product, a decisive factor is to retain the optimal parameters of the extrusion process and immediately cool the extrudate to ambient temperature [3]. The maximum permissible temperature difference is about 15 °C. This prevents condensation of water vapor and the development of the phenomenon of secondary moistening of the product surface. A properly processed full-fat soy product can be stored for at least three months.

Using a baro-thermal treatment of full-fat soybeans, it is possible to obtain a product which is differentiated in terms of physical and chemical features as well as usability. It can be used as a valuable feed component for both monogastric animals and for ruminants.

Similarly, satisfactory results were obtained in the enrichment of the seeds of rape, sunflower and cotton [1, 10].

The extrusion-cooking of legume seeds allowed, in a significant way, the use of these materials in the food and feed industry. This is confirmed by numerous examples of products present on shop shelves, for example, pea extrudates in ready-made foodstuff. In the literature can be found many reports of the essential studies on the enrichment of vegetable materials by extrusion-cooking [2, 10–23].

8.3
Production of Textured Proteins and Meat Analogues

In many regions of the world, the most popular extruded products are textured proteins (TVP) obtained mainly from defatted soy flour (about 50% of protein, PDI 60–90%). In Europe soy steaks, cutlets and other meat analogues have proved very successful in the market owing to their fibrous structure and nutritional value, similar to that of meat products. Such products are in demand with vegetarians but are also highly recommended by dieticians.

Textured proteins of vegetable or animal origin can be processed into meat-like extruded compounds by two methods: dry and/or wet texturization. Dry expanded products are characterized by a spongy texture, they are usually dried and rehydrated for final use (targeted water absorption 2–3.5). Wet extruded products are processed near the final moisture content and therefore do not need to be expanded for more water absorption. They generally have a more fibrous, less expanded texture than dry extruded. Dry textured soy proteins are used by human food producers as meat extenders and analogues because of their linear fibers and firm texture. Dry expanded chunks made from a combination of meat and vegetable proteins are applicable for canned petfoods; a chewy wet cat food is also produced from meat by-products.

As mentioned above, most TVP meat analogues are made from defatted soy and processed primarily by extrusion-cooking. The average moisture content in the process is around 30% and the final composition is extruded at approximately 170–190 °C. Products expand rapidly upon emerging from the die and lose a lot of water by evaporation before they are dried to safe moisture levels, around 7–10%, for storing. Meat analogues must be rehydrated with water or flavoring liquids. A spongy structure produces products with poor flavor retention and lack of real fibrous texture.

Figure 8.2 Twin-screw extruder EVLT 145 (permission of Clextral).

TVP are processed in specially adapted food extruders with L/D > 20 (single or twin-screw) equipped with properly configured plasticization units and pre-conditioners (see Figure 8.2). The processing of defatted soy flour is fairly difficult and requires compliance with strict parameters of baro-thermal treatment. The adjustment of these parameters gives the desired physical characteristics of soy extrudates, which fulfill the role of meat analogues or extenders.

Preconditioning has a substantial role in the production of meat analogues. It is ideal for processing dry TVP, allows replacement of approximately 40% of the mechanical energy by thermal energy, and allows better pre-hydration of larger particles. Nevertheless, there are some negative effects: heating by live steam injection increases moisture by 8–15%, can develop formation of lumps that build-up on the paddles and shafts, especially with wheat gluten, soy white flakes and fresh meats. Preconditioning has only limited use for wet texturized products.

The set of machinery needed to perform dry texturization of TVP is very simple and consists of: a mixer, extruder with pre-conditioner, drying unit and, optionally, a shredder. The basic concept of the processing line is practically slightly extended in relation to the set illustrated in Figure 8.1. The interaction of friction, temperature, shear and moisture must lead the processed raw material to the state of plasticization and gradual denaturation of protein. Properly located mixing screws' elements maintain continuous motion of the mass, which protects it from structural changes before it leaves the extruder's die. While passing through the die, the product (Figure 8.3) should acquire fiber characteristics. Simultaneously, steam evaporating rapidly from the extrudate affects its porosity. The length of fiber of textured protein depends on the inclusions of starch or fiber granules. Their presence clearly shapes the quality of the product [16].

Figure 8.3 Dry textured soy meat analogues.

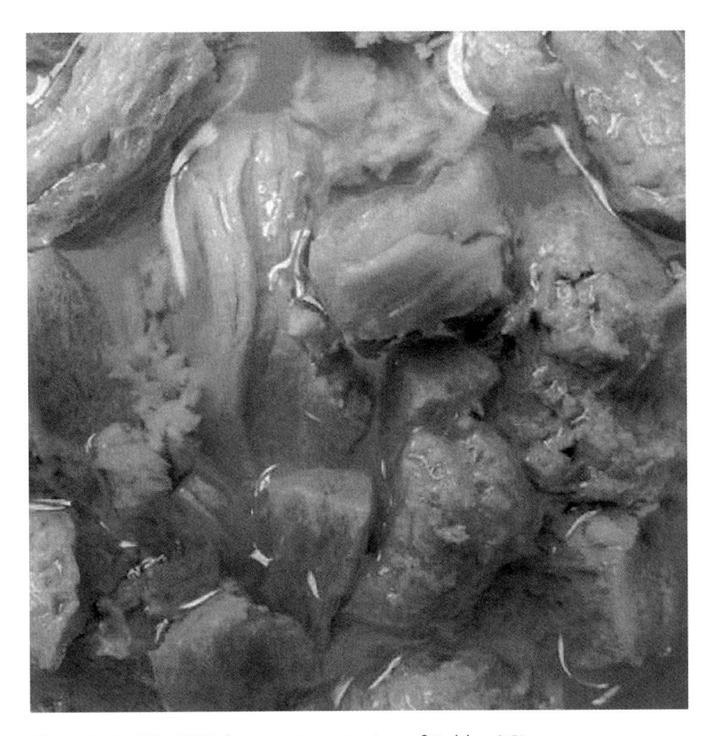

Figure 8.4 Wet TVP for pets (permission of Buhler AG).

Recently, wet texturization has emerged using twin-screw extrusion in a combination of a series of chemical and physical processes to produce a more fibrous structure and meat-like texture of the resulting products (see Figure 8.4). An appropriate pre-treatment allows the use of a larger spectrum of proteins and other ingredients such as fresh meat, fish, starches, fibers and additives. High-moisture texturized proteins are usually processed and packaged in wet condition (pouches, cans or frozen). Due to the more complex technological processes, the extruded mass is subject to an additional treatment in the so-called forming units (additional plasticizing devices), where it is chilled, unified and molded into strips [14]. This method is also applied in the processing of a casein.

References

1 Harper, J.M., Cummings, D.J., Kellerby, J.D., Tribelhorn, R.E., Jansen, G.R., and Maga, J.A. (1978) Evaluation of Low-cost Extrusion Cookers for Use in LDCs. (Annual Report), Colorado State University, July.

2 Matyka, S., Mościcki, L., Jaśkiewicz, T., and Pokorny, J. (1996) Extrusion-cooked faba bean-rapeseed concentrate in broiler chicks diets. *Sci. Agric. Bohem.*, **27** (3), 211–220.

3 Mościcki, L., Mitrus, M., and Wojtowicz, A. (2007) *Technika ekstruzji w przetwórstwie rolno-spożywczym (in Polish)*, PWRiL, Warsaw.

4 Mościcki, L. (1980) Badania nad procesem teksturyzacji soi metodą ekstruzji (in Polish). *Roczniki Nauk. Rolniczych*, **t. 74-C-4**, 47–52.

5 Perilla, N.S., Cruz, M.P., De Belalcazar, F., and Diaz, G.J. (1997) Effect of temperature of wet extrusion on the nutritional value of full- fat soybeans for broiler chickens. *Br. Poultry Sci.*, **38**, 412–416.

6 Wijeratne, J. (2000) The process of dry extrusion. *Feed Technol.*, **4** (2), 10–12.

7 Said, N. (1995) Extrusion processing of ingredients and feed. International Symposium of Feed Production. Curitiba, Brazil, 16 pp.

8 Clarke, E. and Wiseman, J. (1999) Extrusion temperature impairs trypsin inhibitor activity in soybean meal. *Feed Technol.*, **3** (8), 29–31.

9 Zarkadas, L.N. and Wiseman, J. (2000) Inclusion of full-fat soybean in piglet diets. Proceedings of the British Society Animal Science Occasional Meeting: The weaner pig, pp. 45–46.

10 Mościcki, L. and Matyka, S. (1994) Extrusion-cooking of rape seeds. 94 International Symposium on New Approaches in the Production of Food Stuffs and Intermediate Products from Cereal Grains and Oil Seeds, 16-19.11.1994, Begin, People Republic of China, Proceedings, pp. 646–651.

11 Farnikova, L., Reblova, Z., Pudil, F., Pokorny, J., Mościcki, L., and van Zuilichem, D.J. (1996) Browning reactions during extrusion-cooking of peas, faba beans and rapeseed. *Biul. Nauk. Przem. Pasz.*, **1**, 15–28.

12 Matyka, S., Mościcki, L., and Jaśkiewicz, T. (1998) Extrusion-cooking of barley with addition of enzymes to improve its feeding value. Proceedings of 13 International Congress on Agricultural Engineering, Rabat, Morocco, 26.02.98, vol. 6, pp. 225–229.

13 Mercier, C., Linko, P., and Harper, J.M. (eds) (1989) *Extrusion Cooking*, American Association of Cereal Chemists Inc., St. Paul, Minnesota.

14 Mościcki, L., Wójcik, S., Plaur, K., and van Zuilichem, D.J. (1984) Extrusion-cooking to improve the animal feed quality of broad beans. *J. Food Eng.*, **3**, 307–316.

15 Mościcki, L., Wójcik, S., Pisarski, R., and Rzedzicki, Z. (1986) Ekstrudowany bobik

jako substytut soi w żywieniu prosiąt (in Polish). *Biul. Inf. Przem. Pasz.*, **4**, 21–27.

16 Mościcki, L., Wideński, K., and Wójcik, S. (1987) Efekty zastąpienia poekstrakcyjnej śruty sojowej śrutą z nasion łubinu żółtego w żywieniu prosiąt (in Polish). *Biul. Inf. Przem. Pasz.*, **4**, 33–39.

17 Mościcki, L., Pisarski, R., and Wójcik, S. (1988) Użyteczność mieszanek paszowych o różnym udziale ekstrudowanych nasion łubinu i bobiku w żywieniu kurcząt broilerów (in Polish). *Biul. Inf. Przem. Pasz.*, **4**, 22–30.

18 Mościcki, L., Wideński, K., and Rzedzicki, Z. (1990) The research on white lupine extrudate introduction to piglets feed World Review of Animal Production, vol. XXV, 4.

19 Mościcki, L., Tarkowski, A., and Dzirba, L. (1991) Ekstruderat bobikowo-sidowy jako pasza dla bydła (in Polish). *Biul. Inf. Przem. Pasz.*, **2**, 13–22.

20 Mościcki, L. and Matyka, S. (1994) Application of extrusion-cooking for feed premixes stabilization. 6th International Congress on Engineering and Food, Tokyo, 23-27.05.'93, Proceedings of 6th International Congress on Eng. & Food, pp. 879–881.

21 Mościcki, L., Pokorny, J., Kozłowska, H., and van Zuilichem, D.J. (2003) Expander-cooking of rapeseed - faba bean mixture. *EJPAU*, **6** (2).

22 Reblova, Z., Nguyen, H., Pudil, F., Dostalova, J., Pokorny, J., Mościcki, L., and van Zuilichem, D.J. (1996) Nonenzymic browning reactions during the Extrusion and expansion cooking process. International Conference on Chemical Reactions in Food EChS, Prague, Proceedings.

23 Reblova, Z., Piscacova, J., Pokorny, J., Mościcki, L., and Matyka, S. (1995) Changes in glucosinolates and phenolics during extrusion-cooking of rapeseed-legume mixtures. *Sci. Agric. Bohem.*, **26**, 93–104.

24 Thomason, D.M. (1987) Review of processing systems for full- fat soy. En: Full fat soya. A regional Conference. American Soybean Association. Milan, Italy, pp. 114–122.

9
Extrusion Technique in Confectionery

Leszek Mościcki

9.1
Introduction

The use of the extrusion technique in the confectionery industry is a relatively new idea which has found application not only in intermediate operations (pre-processing of components) but also in the manufacture of finished products, for example, spongy sweets and chewing gum (see Figure 9.1). An extrusion-cooker, often referred to as a bioreactor, allows a thermal and baro treatment of raw materials in a controlled and, if necessary, step-by-step manner. In this way, it is possible to affect the physical and chemical properties of processed materials and largely shorten the technological process. So far, the process has been conducted periodically in tubs or tanks operating under atmospheric pressure. By selecting proper extrusion-cooking conditions, the operator is able to control the Maillard reactions, influence the scope of protein and starch transition during emulsifying and gelatinization, and also control the crystallization of sugars, polymorphism of fats and regulate enzyme changes occurring in the processed material. These features have largely determined the success of food extruders among confectionery producers in recent years [1]. Because these are expensive devices their proper selection and knowledge of setting up the plasticization units is of the utmost importance. Only then will their purchase prove economically reasonable and their application effective.

Unlike other sectors of the food industry, the manufacture of confectionery requires particularly careful and labor-intensive preparation of raw materials and a multi-stage production cycle. In most cases, there are a few consecutive and different processing activities and a diversity of raw materials involved: wet, dry, paste, raw or pre-cooked. All components must be pre-mixed and thermally processed before delivery to the last stage of production decisive for the ultimate form of the finished product. A multi-stage conditioning of raw materials plays probably the most important role in the production of confectionery.

Food extruders equipped with an elaborate conditioning system can replace multiple conventional unit processes. This is possible through gradual feeding of

Extrusion-Cooking Techniques: Applications, Theory and Sustainability. Edited by Leszek Moscicki

ISBN: 978-3-527-32888-8

water or steam in various phases of the treatment and varied effects of pressure and shear stresses caused by the configuration of the screws. The selection of the appropriate configuration of the screws, their rotation and the temperature in different sections of the plasticization unit makes it possible to obtain the appropriate product quality in terms of density, chemical properties, taste and shape. All this is achievable in a relatively simple and cost-effective manner in a single device [1–5, 8–22].

9.2 Sweets and Candy

The application of a backward twin-screw extrusion-cooker has much simplified the production of elastic and spongy forms of confectionery products. Well-known and popular, especially among children, flexible sticks, multicolored plaits made from sugar, starch and gelatin (commonly called licorice) today are manufactured mainly on extrusion-cookers (see Figure 9.2). Such a production requires a plasticization unit of $L/D > 25$ [1].

In the first stage of production, the mixture is subjected to intense heat treatment initiating moderate structural changes of starch. Gelatin is pumped, post-starch reaction, into the end zone of the plasticization unit of the extruder, which is intensely cooled. This is to avoid degradation and to consolidate the flexibility of the extruded mass. The extruder's die head is also strongly cooled.

The production of a tough rubbery licorice (Figure 9.3) also requires the use of three basic ingredients: wheat flour up to 43%, sucrose up to 44.5% and water fed

Figure 9.1 Examples of popular extrusion-cooked confectionery products.

Figure 9.2 Examples of elastic sweets.

directly into the barrel. By performing a multi-stage extrusion-cooking at a temperature from 100 to 180 °C, using a plasticization unit with L/D = 40 and a specially set up configuration of screws providing: transportation, material compaction, double kneading (hammers, stars) and degassing; an elastic in nature product can be

Figure 9.3 An example of a tough rubbery licorice.

obtained in the form of strips or spaghetti-like shape of different thickness, requiring no stoving to final moisture content [2, 3, 8].

The application of extrusion-cookers for the production of popular toffees can save up to 35% electric energy. However, this is not the most important factor in this respect; the most appreciated fact about these machines is their ability to maintain stability and repeatable quality of the finished products [1]. While many producers of soft toffees still use single-screw extruders, concurrent twin-screw extruders are gaining increasing popularity due to their ability to control and adjust the mixing of the mass. Well-chosen, self-wiping screw elements eliminate the threat of material combustion, the coalescence of fat and rapid changes in the texture of a mass. Too intense a mixing may affect the re-orientation of protein to a fibrous form, which is not a favorable phenomenon. Oil and flavor additives are applied in the final section of the barrel.

The traditional manufacture technology of toffees is associated with the development of large quantities of reducing sugars in the processed mass. In the recipes of extruded toffee mixtures the sucrose : glucose ratio should equal 1 : 1. Huber [4] recommends that in this case malt sugar be fed into the extruder as a liquid premix.

Food extruders find application in the production of caramel and evaporated products characterized by specific rheological and functional features (degree of saccharification, crystallization, etc.). Extrusion-cookers help control the rheological transition of the mass during its processing and maintain an amorphous structure of the final product. In such a case, of vital importance is a properly planned multi-stage heat treatment that involves not only the heating of the mass but also its cooling at the most appropriate moment [5–9].

The use of the co-extrusion technique can offer many different forms of filled sweets. This method is used in the manufacture of pretzel-like candy formed of "single crystal" tubes filled with air or a nut filling. During their production, two basic conditions must be observed:

- a relatively short residence time of the material inside the extruder in order to avoid inversion and excessive adhesiveness of the mass;
- efficient degassing during extrusion, which guarantees increased transparency of the product and thus enhances the visual appearance of products.

9.3
Creams and Pastes

The production of hard and soft cream pastes requires special procedures that guarantee sufficient consistency of the emulsified and wet mass containing numerous small crystals. It is a laborious and time-consuming activity. The use of an extrusion-cooker largely simplifies the technological process owing to the option of direct feeding of components to the device. Ground dried fruit, nuts, sugar, fat, heated and dissolved gelatin, glycerin or sorbitol are fed into the barrel in specific phases of the processing, virtually in a monocycle. By adjusting the appropriate extrusion temperature, the operator is able to eliminate the effect of "dry skin" on the surface of the product. To diminish the occurrence of excessive crystallization of sugars during extrusion, it is necessary to reduce the content of glucose syrup in a material mixture to a maximum of 20% [1, 2].

As commonly known, the traditional methods of production of creams unfortunately involve a risk of bacteriological contamination, especially during the preparation of raw materials in a non-concentrated form. In the case of extrusion-cooking, dry and powdered components such as gelatin or arabic gum can be fed directly to the processed mass in the extruder, where they are dissolved in a small amount of water at high temperature [2, 6, 10]. The excess of humidity is removed in the extruder's degassing unit.

Flavor additives are supplied under pressure in the end part of the extruder, in the final phase of extrusion. Of great importance is intensive cooling of the extrudate immediately on leaving the die. This helps maintain taste and portion the mass.

The application of an auxiliary co-extrusion equipment or a die for the making of sheet (as in the production of snack pellets) offers filled or multi-layer products, coated with chocolate or sugar icing at a later stage of production.

9.4 Gums and Jellies

The origin of the production of chewing gum is derived directly from the processing of plastics. Chewing gums as opposed to jellies exhibit higher viscosity and flexibility in a hydrous environment (the mouth). They should not stick to the teeth and, of course, not dissolve (soluble gum is produced by other methods from traditional confectionery masses with the appropriate additions).

Chewing gum is often made in plastic extruders. Much better results are obtained, however, using specially prepared food extruders. Why is that? They are more useful in the manufacture of various forms of products and ensure better production stability, lower energy consumption and need fewer personnel [1, 12, 14, 17, 20]. Furthermore, extrusion-cooked gum retains flavor and flexibility for a longer time, despite intense chewing; it is also more durable. Extrusion-cooked chewing gum of mint flavor maintains its fresh taste while chewing for 50% longer than conventionally made chewing gum.

Unlike other confectionery extrudates, in the production of chewing gum taste additives are applied at the very beginning of the baro-thermal treatment. This is done to ensure their maximum dispersion in the processed mass. A similar approach has been adopted for cocoa, coffee, fruit powders, citric acid and special additives, such as anti-smoking products for smokers or teeth whitening substances. Sometimes it is also necessary to add chemical components to reduce the product viscosity. All these ingredients are fed in appropriate proportions.

An extrusion-cooker for the production of chewing gum must be equipped with properly configured screws and additional peripherals (Figure 9.4). The working unit consists of two sections:

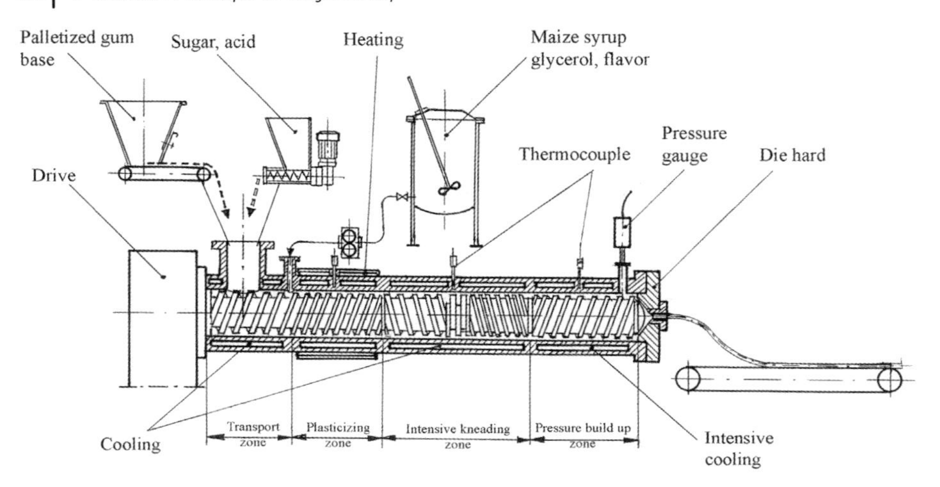

Figure 9.4 The extrusion of chewing gum.

- A short section, heating the mass to around 70 °C; here plasticization takes place as well as absorption of sugars and taste additives;
- A mixing-forming section; here at a temperature below 50 °C the final kneading of dough and product forming takes place.

The production of chewing gum is not as simple a process as it may seem and requires careful attention from the operator. What is more, failure to maintain a technological regime, poor selection of raw materials and, above all, the inappropriate choice of extruder and its fittings may very soon thwart the ambitions of a manufacturer. Non-compliance with the optimal processing parameters can easily lead to starch dextrination, which will certainly affect the extrudate viscosity. Thermal processing must be very precise, smooth and kept to a minimum so as not to destabilize the texture and lead to the deformation of the final product. Positive results in the extrusion of chewing gum are obtained when modified starch is used, enabling the manufacture of gum of delicate consistency. The substitution of glucose with sucrose positive influences the extrudate's viscosity and adhesiveness.

Technologists know many methods and techniques for obtaining high-quality chewing gum. Moreover, of utmost significance is their expertise and experience.

The production of popular jellies (Figure 9.5) is less complicated than the manufacture of chewing gum. The finished product is more susceptible to formation and contains higher moisture [1, 17, 20]. This does not mean that the excess of water makes no difference; quite the contrary, the use of water should be limited to a minimum so as to maintain an adequate consistency of the extrudate. A processing diagram is shown in Figure 9.6.

The mixture of agar, gelatin, pectin, syrup and glucose is supplied to the first section of a single-screw extrusion-cooker, where it is subjected to thermal treatment at a temperature below 145 °C. The second section handles degassing with a vacuum pump and later cooling and forming at a temperature not exceeding 80 °C.

Figure 9.5 Popular jellies.

Starchy jellies should display a high degree of gelatinization, that is, reach an approximately 95% isotopic form observed in a refractometer as a uniform mass [2, 14]. In this form, starch granules are swollen and very flexible. Starch mass of low viscosity requires more mechanical energy during mixing and this is converted into

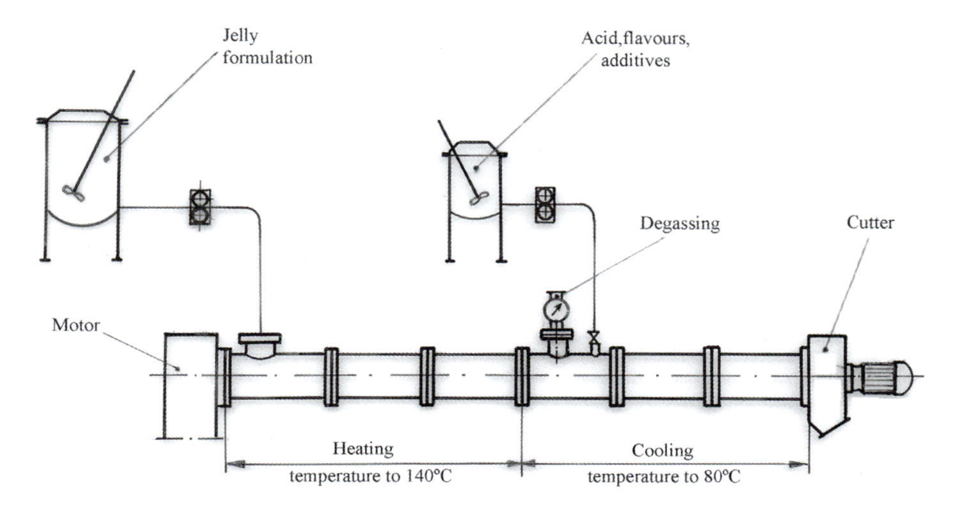

Figure 9.6 Production of jellies.

heat. If the grains crack, amylopectin transforms into a uniform mass of high viscosity. To prevent this, while configuring the screws, it is necessary to replace the elements increasing shear rate by transporting components in the location where this phenomenon is expected to occur.

High temperature and the presence of shear stresses can cause degradation of amylose and amylopectin to short-chain forms. Furthermore, during the extrusion-cooking of jellies from mixtures of low humidity, the gel boiling point may be easily exceeded, leading to higher pressure of extrusion and a decline in its performance. In order to overcome this obstacle, it is advisable to inject steam directly into the barrel. This will increase the efficiency of the extruder but will not significantly raise the moisture content of the extrudate (excess water will evaporate after leaving the die).

9.5 Other Products

Wiedmann and Rapp [23] proposed the possibility of continual production of chocolate through multi-stage extrusion on twin-screw food extruders by Werner&Pfleiderer. Clextral also developed a method for the production of chocolate using their own food extruders, which, in the opinion of French experts, reduces the manufacture time, labor costs and the number of auxiliary production facilities. It also allows the manufacturer to narrow the production area [1]. These alternative forms of chocolate processing are still moderately acknowledged, but may appear as good prospects for the future.

Aerated confections constitute a complementary offer of confectionery products available in stores. Formerly, the production of this type of product required manual labor involving sieves, rolling, cutting and drying of the mass. Today, they are manufactured by high-pressure kneading extruders which facilitate the absorption of gases in fats. Additionally, manufacturers inject air or CO_2 in the final stage of extrusion. These kneading elements of screws are also used in the manufacture of products such as nougat [11, 22].

In order to fix the porous form of extrudate, it is necessary to apply strong cooling to the extruder's die down to a temperature of no more than 90 °C. The confectionery extrudates obtained exhibit a density of less than $0.15\ \mathrm{g\,cm^{-3}}$. The die head may be rotated to give different colored strands as a rope (twisted and porous products at the same time maintaining entire control of the production process). The products of this kind are very attractive to children and may exhibit numerous sensory features, shapes and colours.

Food extruders producing toffee and jellies can be set up with an ice-cream machine making filling for co-extruded forms of frozen sweets [1, 2, 10]. The creativity in the adjustment of die nozzles of extruders, the selection of ingredients, flavors and colors offers amazing results. Ice-coated or filled extrudates, cut and packed in attractive packaging comprise the showpiece of the latest generation of frozen confectionery products.

9.6 Concluding Remarks

The production of compound bars, pads or filled cookies belongs to the field of co-extrusion which is used in many branches of the agri-food industry. Today's high-performance food extruders successfully compete with the conventional production methods applied to many confectionery products [20, 24]. It is not an overstatement to say that the implementation of the extrusion-cooking technique has enabled the manufacture of many hitherto unknown products that have gained widespread consumer recognition. This new field of confectionery processing is developing very rapidly and has an unlimited potential, as seen through a surge in the quantity of new extrudates presented by the confectionery industry.

References

1 Best, E.T. (1994) Confectionery extrusion, in *The Technology of Extrusion Cooking*, (ed. N.D. Frame), Blackie Academic and Professional, London.

2 Guy, R. (2001) *Extrusion Cooking. Technologies and Applications*, CRC Press Inc., Boca Raton, FL.

3 Van Zuilichem, D.J., Tempal, W.J., Stolp, W., and van Riet, K. (1985) Production of high boiled sugar confectionery by extrusion cooking of sucrose:liquid glucose mixtures. *J. Food Eng.*, **4** (1), 37–51.

4 Huber, G.R. (1984) New extrusion technology for confectionery products. *The Manufacturing Confectioner*, **64** (5), 51–54.

5 Fletcher, S.I. and Jones, S.A. (1987) Engineering Aspects of the Extrusion Cooking of Confectionery. *Leatherhead Food RA Scientific & Technical Survey*, **160**, 21–22.

6 Lacey, R.W. (1991) *Unfit for Human Consumption*, Souvenir Press, London, pp. 190–202.

7 Rapaille, A. (1990) New technological trends for the production of starch gums and jellies. *Confectionery Manufacture and Marketing*, **27** (6), 51–52.

8 Rice, P. Continuous extrusion cooking – a new technique in liquorice manufacture. *Confectionery Manufacture and Marketing*, **21** (5), 21–22.

9 Vincent, M.W. (1984) Extruded confectionery, equipment and process. *Confectionery Manufacture and Marketing*, **21** (11), 34–36.

10 Jackson, E.B. (1987) Glucose Syrups and Starches, their Types and Applications for Cooker Extruder Products. Paper presented at Zentralfachschule der Deutschen Sußwarenwirtschaft, Solingen Germany.

11 Treiber, A. (1984) The Buss confectionery process for pressed tablets, vermicelli, nonpareils etc. on the Buss kneading-extruder. *Confectionery Manufacture and Marketing*, **22** (4), 6–7.

12 Zallie, J. (1988) New starches for gelling and non-gelling applications. *The Manufacturing Confectioner*, **68** (11), 99–104.

13 Reidel, H.R. (1980) Aerated sweets and bars produced on special extruders and vacuum expanders. *Confectionery Production*, **46** (2), 64–66.

14 Rapaille, A. (1985) Practical aspects of the production of starch gums by extrusion. Proceedings of Koch und Extrusiontechniken, Zentralfachschule der Deutschen Sußwarenwirtschaft, Solingen-Gräfrath, Germany.

15 Wright, S.J.C. and Dobson, A.G. (1984) Novel Methods for the Manufacturing of Confectionery Products, Part II, Injection Moulding of Gelatine Gums, an initial study. Research Report no 484, Leatherhead Food RA, Surrey, England.

16 Daquino, A.J. *et al.* (1985) Extrusion cooking process simplifies candy making. *Candy Ind.*, **11**, 38–44.

17 Elsner, G. and Wiedmann, W. (1985) Cooker extruder for the production of gums and jelly articles. *Impulse Foods Suppl.*, **2**, Nov.

18 Ganzeveld, K.J. and Janssen, L.P.B.M. (1991) *Extrusion Commun.*, **4** (2), 13.

19 McMaster, T.J., Smith, A.C., and Richmond, P. (1988) Physical and rheological characterization of a confectionery product. *J. Texture Stud.*, **18** (4), 319–334.

20 Mercier, C., Linko, P., and Harper, J.M. (1989) *Extrusion Cooking*, American Association of Cereal Chemists Inc., St. Paul, Minnesota, USA.

21 Renz, K.H. (1987) Extrusion, does chewing gum pass the taste test. *Food Manufacture Int.*, **4** (4), 35. 37. 39.

22 Vessa, J.A. (1991) Confectionery processing on a kneading extruder. *The Manufacturing Confectioner*, **71** (6), 83–86.

23 Wiedmann, W. and Rapp, R. (1985) The 30 minute chocolate process. Proceedings of Chocolate technology 85 Seminar, Zentralfachschule der Deutschen Sußwarenwirtschaft, Solingen-Gräfrath, Germany.

24 Wiedmann, W. and Strobel, E. (1987) Processing and economic advantages of extrusion cooking in comparison with conventional processes in the food industry, in *Extrusion Technology for the Food Industry* (ed. C. O'Connor), Elsevier Applied Science, London, pp. 132–169.

10
Pet Food and Aquafeed

Leszek Mościcki

10.1
Introduction

The devices which have pioneered the idea of extrusion-cooking in feed processing were modified single-screw extruders by Anderson. They were used for heat treatment of oilseed cake. Currently, these machines have been substantially upgraded and are widely used in the manufacture of pet foods and aquafeed. The modification of the screws in extrusion-cookers increased the mixing capacity, the influence of shear and temperature generated as a result of strong internal friction of extruded mass. In this way, large extruder throughput can be achieved, up to 15 tonnes per hour, and far more thorough transformation of the processed raw material [1–4]. The increasing demands of feed manufacturers, who discovered the huge potential in extruders, are stimulatng further improvement of these machines. Initially, manufacturers used only single-screw extruders; today, twin-screw co-rotating extruders are becoming commonplace offering modular structure and varied geometry of the working elements [7, 10].

In the feed industry, extrusion can compete with pelleting only where the thermal treatment of raw materials is needed (deeper chemical transition) and the agglomeration of feed of higher moisture content (>30%).

The energy consumption of feed extrusion is approximately 0.1 kWh/kg of product [3]. This relatively high energy input has caused the extrusion-cooking technique to be used chiefly to produce specialized feed, such as pet food and aquafeed, as well as feed for young, breeding stock. An important factor is that extrusion-cooking allows full or partial inactivation of antinutritional components in the processed raw materials and also increases their digestibility, which makes them especially suitable for young animals.

Extrusion-Cooking Techniques: Applications, Theory and Sustainability. Edited by Leszek Moscicki

ISBN: 978-3-527-32888-8

Figure 10.1 Various forms of extrusion-cooked feed.

10.2 Market Development

In the EU countries, the extrusion method is employed for the manufacture of large quantities of specialized pet food, and food for breeding stock and fish (see Figure 10.1). A dynamic growth of this sector has been observed for more than 15 years. This also is the case in Eastern Europe, especially when it comes to the sales of feed for cats and dogs [5]. The indisputable advantages of extruded feed, attractive ways of promotion and a huge potential in terms of the number of potential customers (Poland tops the European rankings) have caused an immense increase in the production and sales of this type of product in that part of the world. It should be also noted that the production of extruded feed is highly profitable.

Figures 10.2–10.6 show demographic and economic indicators for the pet food industry. Upon analysis, all indicators prove a dynamic development of this sector.

10.3 Feed Extruders

As already mentioned, as with food processing and the production of simple snacks, so with the feed industry which also occasionally uses autogenic single-screw extrusion-cookers. This is certainly due to the intention to minimize investment expenditure [4, 6]. Autogenic units, which produce heat from the friction of material

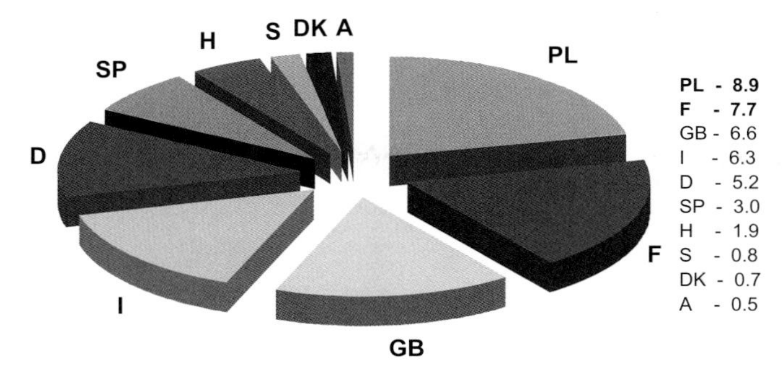

Figure 10.2 The population of dogs in Europe (in millions) [5].

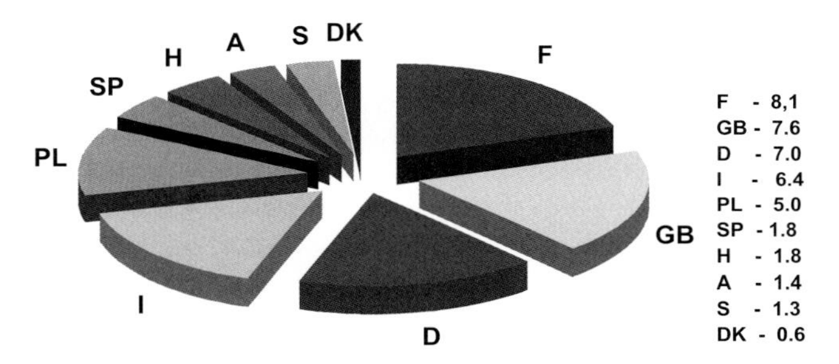

Figure 10.3 The population of cats in Europe (in millions) [5].

being forced through the barrel, are generally used for the production of extrusion-cooked feed components. They can also be employed to produce pet food with a simple recipe. The process temperature in these devices can be adjusted to a minor extent by changing the size of the partitions in the barrel depending on the type of

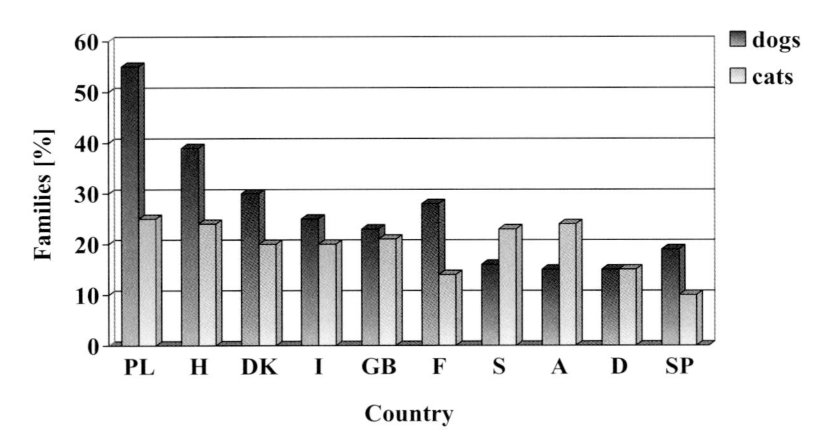

Figure 10.4 Families with pets in Europe [5].

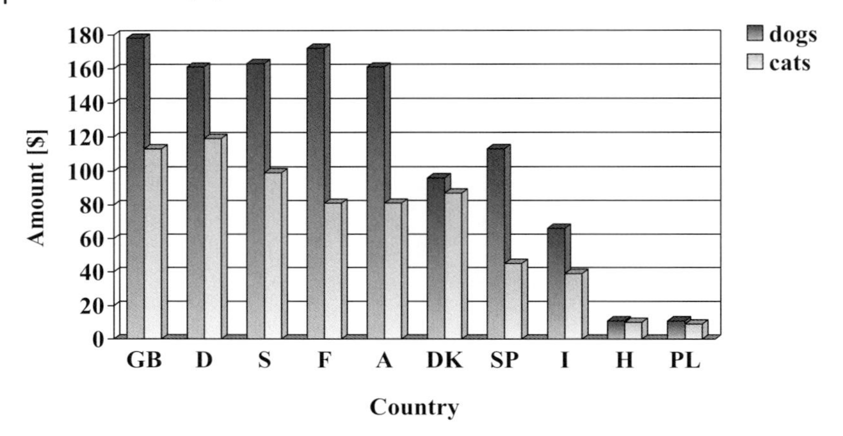

Figure 10.5 Average annual expenditure on pet food in Europe (in US dollars) [3].

processed material. Despite the simple construction of these devices they can provide [1, 3, 6, 7]:

- cleavage of cell membranes in the processed material,
- starch gelatinization,
- destruction of antinutritional ingredients,
- raising of the quality of raw materials, such as extrusion-cooked sunflower or soya meal,
- improvement in the digestibility of starch and protein,
- increase in durability (e.g., in the case of rice bran),
- sterilization (destruction of salmonella).

Single-screw feed extruders of more complex design have a more diverse application. They are equipped with an adjustable heating–cooling system and different plasticization units.

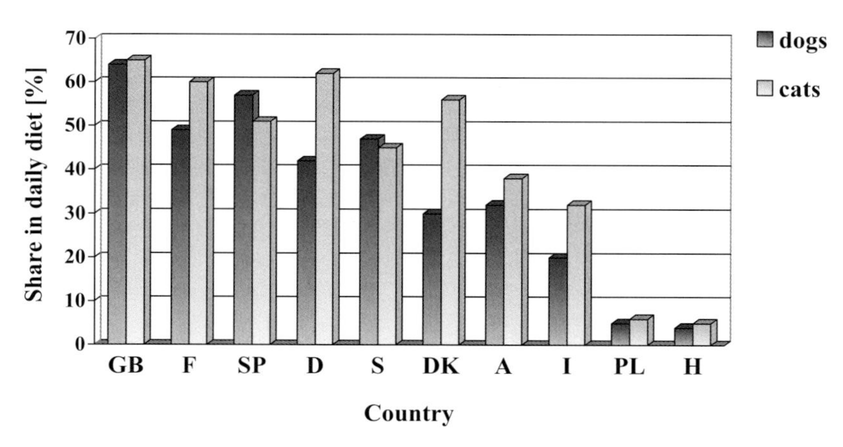

Figure 10.6 Percentage share of manufactured pet food in pets' daily diet [3].

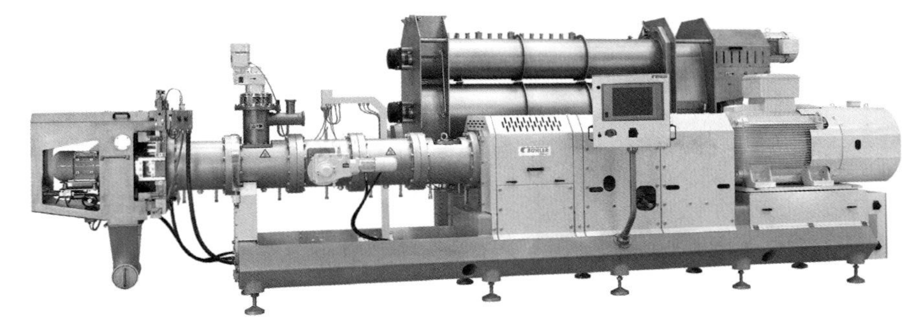

Figure 10.7 Modern twin-screw feed extruder, type Ecotwin (permission of Buhler AG).

In recent years, single-screw feed extruders have been supplanted by more expensive but more energy-effective twin-screw extruders (Figure 10.7). The main advantages of contemporary twin-screw feed extruders are [8, 10]:

- versatility in the processing of raw materials, possibility to obtain a variety of extrudate shapes and processing of raw materials with higher moisture content,
- perfect control of all the parameters of the extrusion-cooking process,
- the option of regulating (creating) extrusion pressure in the barrel, and, consequently, of extrudate density,
- perfect mixing of components,
- capacity to produce new, attractive products of sophisticated form,
- quick readiness to perform in stable working conditions.

It is therefore not surprising that the production of feed by means of twin-screw extruders has recently won more and more supporters. Their application is becoming more cost-effective because the market offers more energy-efficient, high-performance machinery.

When it comes to the production of extrusion-cooked feed, there are two principles which should be closely complied with in order to achieve the desired production effects [1, 3]:

- protein materials or oilseeds should be processed on long-barrel extruders with $L/D > 16$, equipped with conditioners and screws of specially selected geometry;
- starch materials should be subjected to baro-thermal treatment in extruders equipped with shorter plasticization units with $L/D = 6–12$.

10.4 Technology

Dry expanded pet foods, nutritionally balanced for the daily requirements of the animal, are the most commonly produced extruded diets. Worldwide, more than 80% of all dry pet foods are produced on single-screw extruders. As dogs are animals

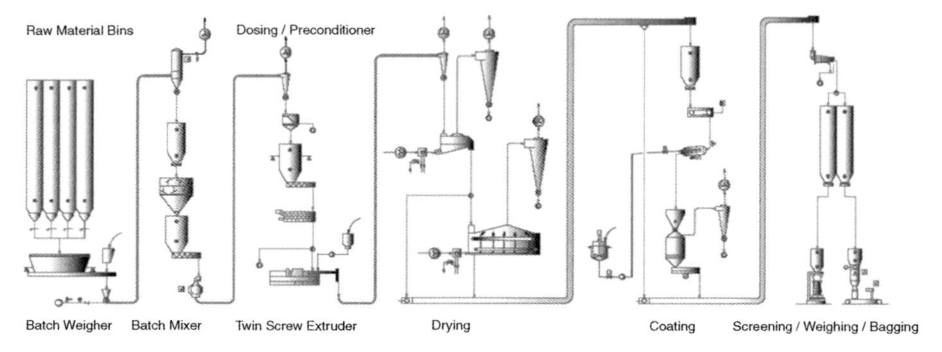

Figure 10.8 A production line for petfood and aquafeed [4].

with omnivorous feeding behavior their formulations contain a significant amount of cereals with a good portion of animal proteins. The total protein content varies, generally in the range of 22–28% protein with a fat content around 8–16%. Cats are distinctly carnivorous animals and require higher protein contents of 28–34% with fat contents around 15–20% and significantly lower carbohydrate content. They prefer fresh meat addition and do not like reaction flavors and ingredients over-exposed to temperature. Such expanded products are often produced in a variety of shapes, textures and colors to suit the size and the dentition of the animal, as well as to appeal to the potential buyer [8, 11, 13].

A set of standard manufacturing equipment comprising the basis of a processing line is shown in Figure 10.8.

According to estimates, the total cost of production of 1 tonne of extruded pet food or aquafeed on the presented line ranges from €13 to €30. In the developed world, extruded aquafeed comprises 100% of the feed used in breeding farms, because it exhibits positive reaction to an aqueous environment and superb nutritional features. Fortunately, such feed is also beginning to attract the growing interest of fish farmers in developing countries.

Table 10.1 summarizes the estimated energy consumption and water during the production of standard aquafeed for trout fish on an extended processing line with a capacity of 12 $t\,h^{-1}$. Examination of these data will help you prepare the necessary balance of expenditure related to such production.

10.4.1
Raw Materials and their Preparation

As regards the production of popular pet food and aquafeed, the material used is cereal grains together with their milling waste, post-extraction soya meal, animal meal, powdered milk, animal fats, vegetable oils and premixes (vitamin and mineral supplements). Raw materials must be ground to particles of size 0.15–0.4 mm. For the production of aquafeed, much finer grinding is recommended.

Table 10.1 The demand for energy and water during the manufacture of fish feed at the output of 12 t h^{-1} [3].

Demand for electric energy (kW)		
Extruder	210 × 80%	168.8
Drying/cooling	71 × 80%	57.0
Total		225.8
Total kW t^{-1}	225.8/12.5	18.1
Demand for steam (kg h^{-1})		
Extruder		1875
Drying/cooling		3750
Total		5625
Total (kg t^{-1})	5625/12.5	450
Demand for water (kg h^{-1})		
Added water		1250
Cooling water*		2500
Total		3750
Total (kg t^{-1})	3750/12.5	300

*in closed cycle.

Unfortunately, in the course of thermal treatment, although it is the HTST type, there is a loss of vitamins in extruded materials, especially vitamins C and B and, to a lesser extent, vitamins A, D and E; their losses amount to 15–20% [1, 3, 7]. To remedy this condition, it is necessary to add vitamins in the end stage of production, that is, during the coating of extrudates with oil or fat.

Table 10.2 shows a standard recipe for high-quality feed for dogs and eels (floating feed) illustrating the demand for the material and enabling the calculation of production costs.

In the production process, the pre-conditioning of the mixture in a conditioner plays a pivotal role, where steam or water (up to 30% of the dry substance) as well as a few percent of fat are added during mixing. Conditioning excellently facilitates

Table 10.2 Popular recipe for dog and eel feed [3].

Material	Dogs (%)	Eels (%)
Maize	44	—
Soya meal	17	10
Meat and bone meal	17	—
Wheat flour	—	28
Wheat meal	16	5
Fish meal	—	53
Premix	1	2
Fat/oil	5	2

the process of raising of the material and starch gelatinization in no more than a few dozen seconds. Pre-conditioning of the mixture clearly enhances the extruder's productivity.

10.4.2
Extrusion and Drying

Depending on the recipe, purpose and complexity of the form of extrusion-cooked feed, it is necessary to choose the appropriate configuration of screws and the type of extruder. As mentioned earlier, twin-screw extruders of modular design are more universal devices. Unfortunately, they are much more expensive. One of their advantages is the option of addition of fresh meat or its waste to the mixture, which increases the sensory values of pet food and facilitates the agglomeration of extrudate. Having twin-screw extrusion-cookers, the manufacturer is able to produce stuffed products and feed of high fat content as well as very fine forms such as aquarium feed [2].

By appropriately adjusting the parameters of the extrusion process, such as temperature, pressure, residence time, screws' rpm, shape of the die and rotation of the knife, it is possible to shape the properties of the final product. These are first and foremost the texture, the scope of chemical transformations, the density, the degree of expansion, and so on. For example, the feed for carps should sink in the water but maintain its consistency for a specified time. On the other hand, trout fish require floating feed. These properties of feed can be adjusted by appropriately controlling the treatment process [12, 14].

Table 10.3 shows how the shape of the die may influence some physical characteristics of the extrudate.

Rapid development of the production technology allow pet food producers to manufacture very attractive products nowadays [8]. Additional investment in sophisticated equipment makes it possible to achieve extraordinary consistency, shape and appearance of the extrudates – properties of more importance for the pet owners than

Table 10.3 The influence of the shape of the die on the physical properties of extrudate and the motor load (feed production at constant parameters).

Die configuration ∅ 4 mm	Bulk density (kg m^{-3})	Gelatinization (%)	Extruder load (%)	Extrudate appearance
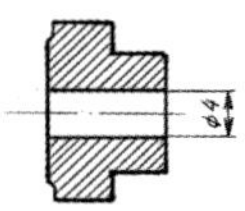 ø4	>310	90	>90	smooth, closed surface; cylindrical shape
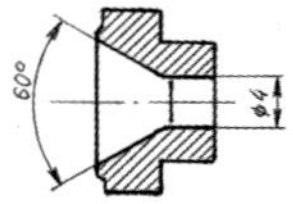 60°, ø4	>340	80	>70	porous surface; spherical shape

for the pets themselves. Below is a list of the most attractive products and their mode of manufacture:

- **Bicolor/Marbled** – the mass flow is separated, differently dyed and reunited in a common die, obtaining extrudates with the same texture and same bi-color pattern.
- **Dual color** – the mass flow is separated, differently dyed and extruded through separate dies, obtaining extrudates with the same texture but different color.
- **Bitexture** – two different formulas processed under different condition are processed through a common die, obtaining extrudate with different texture and/or different color
- **Co-extrusion** – the cooked mass from one extruder is shaped into a hollow profile, while a filling mass is injected, by means of a pump, into the cavity; the filling mass typically is a meltable, fat or sugar-based compound.

Most assortments of extruded feed require a fairly intensive process of drying. In small processing lines, the drying units may be vertical (heated by gas or electricity); however, more popular are high-performance belt dryers with a multi-stage drying process relying on hot air. Most frequently, the amount of water to be removed from the extrudate amounts to 10 to 15%, which needs more than 10 min drying. Certainly, the cracking of extrudates should be avoided, which may become particularly severe in the case of larger products. As a rule, the drying temperature does not exceed 100 °C.

The final stage of production is the application of flavors and vitamins and oiling with spray. After a brief cooling, the product may be packaged.

Some aquafeed or poultry feed requires a much higher fat content in the recipe (i.e., the feed for salmon should have more than 30% fat). High fat content impedes the process of agglomeration during extrusion, and this, in turn, translates into its bonding in the product. A conventional coating of extrudates may also fail. In this case, a solution may be the addition of fat in the end phase of feed production in special vacuum mixers [1, 8–10]. At reduced pressure, porous extrudate much more readily absorbs (sucks) sprayed fat that penetrates deeper into the layers of the product so that its content can be raised to 35–40%; in standard conditions this is unworkable.

The examination of the product quality, besides its organoleptic evaluation, follows the same procedure as adopted for the analysis of typical feed pellets. Much attention is attached to the water absorption index (WAI) of the extrudates and their behavior in an aqueous environment (especially with aquafeed). The nutritional value of the feed is tested *in vitro* and, in the case of comprehensive measurement, by *in vivo* methods.

By application of the extrusion-cooking technique, feed manufacturers have launched a new range of feed products which previously were virtually unknown or produced in a costly and time-consuming manner. The use of many waste materials or materials of no great economic importance has allowed producers to economize on production and contributed to the decline in the consumption of meat and its products by pets. Extracted feed is very efficient, effective and convenient in application and storage (Figures 10.9 and 10.10).

Figure 10.9 Samples of colorful pet food (permission of Buhler AG).

10.5
Concluding Remarks

Extrusion-cooking has become a very useful and economical method for production of very convenient and nutritional feedstuffs for pets and aquatic life. Selecting

Figure 10.10 Aquafeed (permission of Buhler AG).

proper equipment permits better utilization of available vegetable raw materials to produce cost-effective healthy diets with improved and unique feeding characteristics. Petfood production can be very profitable when the desired process management of the raw material formulation is applied. Moreover palatable, functional and tailor-made products can be easily manufactured using ingredients which were previously under-utilized or poorly accepted [1]. That sector is in permanent development in many regions of Europe, where there is still enough room for new producers.

Over the last decade, the world has witnessed spectacular growth in the aquaculture industries, and no doubt this tendency will be continued, especially in the developing countries of Asia and Africa. Nutrition and feeding play a central and essential role in the sustained development of aquaculture and, therefore, "engineered" aquafeed is the most attractive solution for fish farming.

References

1 Frame, N.D. (1994) *The Technology of Extrusion Cooking*, Blackie Academic & Professional, London.

2 Munz, K. (2006) Extrusion technology in the animal feed industry Buhler AG, rev. 18 April.

3 Mościcki, L., Mitrus, M., and Wojtowicz, A. (2007) *Technika ekstruzji w przetwórstwie rolno-spożywczym (in Polish)*, PWRiL, Warsaw.

4 www.buhlergroup.com (2010) Informative and technical materials.

5 Mościcki, M. (1999) Polish petfood and aquafeed market shows great potential. *Feed Technol.*, **3**, 56–57.

6 Hauck, B.W. (1991) An Overview of Food Extrusion Costs. Presented at Extrusion Short Course, University of Nebraska, Lincoln, Nebraska, November 6–11.

7 Mercier, C., Linko, P., and Harper, J.M. (1989) *Extrusion Cooking*, American Association of Cereal Chemists Inc., St. Paul, Minnesota, USA.

8 Munz, K. (2008) Pet food production. printed materials of the Workshop Petfood/Aquafeed, Uzwil, Switzerland, 24. Sept.

9 Andrews, J.W. and Davies, J.M., (1979) Surface coating of fish foods with animal fat and ascorbic acid. *Feedstuffs*, **51**, 33.

10 Guy, R. (2001) *Extrusion Cooking. Technologies and Applications*, CRC Press Inc., Boca Ration, FL.

11 Juśko, S., Walczak, I., Mościcki, L., and Hodara, K. (2001) Wpływ dodatku odpadów drobiowych na proces ekstruzji karmy dla psów (in Polish). *Inżyn. Roln.*, **10**, 179–185.

12 Kazemzadeh, M. (1991) Engineering fish foods. *Extr. Commun.*, **1**, 12–13.

13 Teeter, W.A. (1979) Carbohydrates for dogs. *Pet Industry*, **21**, 25.

14 Wójtowicz, A., Mościcki, L., and Łopacki, T. (2001) Badania właściwości fizycznych ekstrudowanej karmy dla zwierząt domowych (in Polish). *Acta Agrophys.*, **46**, 215–226.

11
Expanders

Leszek Mościcki

11.1
Introduction

In the 1980s, the world market for feed machinery spawned the production of industrial expanders. The idea of their operation is based on the pin extruder (see Chapter 1), where the processed material is further subjected to mixing and heat treatment. The scope of application of these devices increases every year. Initially, they were used for pre-treatment before the pelleting of feed, and now expanders are used in autonomous processes: to increase the nutritional value of full fat soybeans and other pulses, to sterilize feed components, and, in some cases, to produce simple extrusion-cooked feed stuff [1–4].

It is worth mentioning that there are many similarities between the process occurring in the expander and that in the extrusion-cooker, because their origin is baro-thermal treatment realized under slightly different conditions in a machine with different layout of the plasticization unit (screw and barrel). The essential difference between extrusion and expansion is that the latter is less energy-intensive and that, at the exit of the installation, the die is replaced by a conical discharge valve (most popular solution). The expansion process can be applied directly to a foodstuff, or to an individual ingredient, and sometimes is even used as part of a more complex system whereby the raw material is expanded after cooking.

Unlike single screw food extruders, expanders have a simple design and are easier to control. Of certain importance is the economic factor of their application. For example, if an expander is used, the production costs constitute 30% of the production costs generated by a twin-screw food extruder and about 50% of the production costs for a single-screw extruder [2, 3]. Of course they can be used only for limited purposes, therefore are not always applicable.

Table 11.1 shows the main benefits of the expander application versus the extrusion-cookers (single- and twin-screw devices).

Extrusion-Cooking Techniques: Applications, Theory and Sustainability. Edited by Leszek Moscicki

ISBN: 978-3-527-32888-8

Table 11.1 An outline of results for an expander and food extruders in the production of animal feed [4].

Factor	**Single-screw extruder**	**Twin-screw extruder**	**Expander**
Energy consumption (kWh t^{-1})	30–70	40–80	12–30
Pressure (MPa)	1–8	1–10	1–6
Process temperature (°C)	90–160	90–180	80–140
Material m. c. (%H_2O)	18–28	18–60	12–30
Degree of starch gelatinization (%)	80–100	90–100	40–80
Solubility of protein (%)	below 10	below 10	10–12
Agglomeration capacity	yes	yes	no/limited
Mixing effect	poor/average	good	poor
Dependence on material content	average/high	independent	high
Heating option	yes	yes	yes/no
Cooling option	yes/no	yes	no
Self-wiping	no/poor	good	no
Investment expenditure	average/high	high	low/average
Maintenance costs	average/high	high	low/average

11.2 Design of Expanders

The essential components of an expander are the supply and preconditioning units, the barrel complete with a screw and vapor injection valves, the hydraulic system at the die which regulates the pressure level, and the expander shaft powered by a motor. The expander itself is a specially designed single-screw device while the barrel is equipped with mixing or kneading pins and the screw flights interrupted at the pin locations (see Figure 11.1). Usually a screw has a diameter of 150 to 500 mm and they

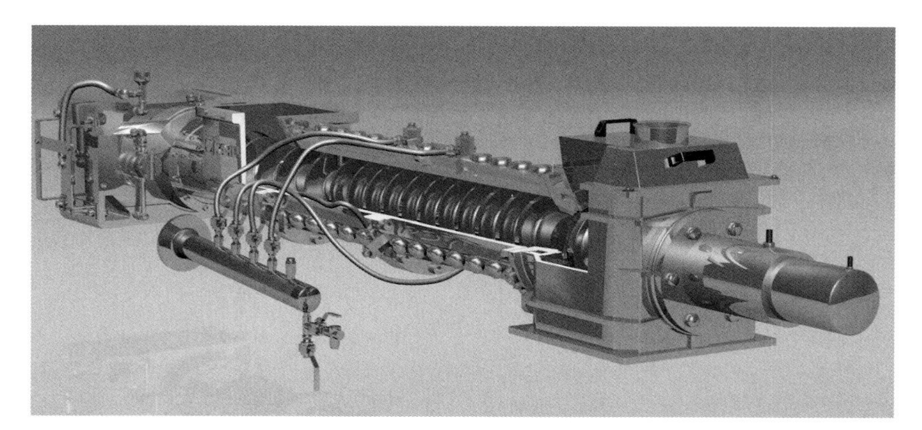

Figure 11.1 An annular gap expander (with permission of A. Kahl GmbH).

Figure 11.2 Various types of expandates.

can perform with a capacity up to 30 t/h. They are equipped with various die-heads: annular gap (conical ring-shaped), finger-shaped, flange-shaped or aperture-disk-like, and a relatively simple plasticization system. Many of these solutions are patented.

The type of die largely determines the product quality and the course of the processing (see Figure 11.2). For the user, the agglomeration capacity is particularly important, followed by the opportunity to shape the product, starch gelatinization, the heat treatment and to what extent these factors will be supportive in the further production process.

The most common die design in expanders is an annular gap ring-shaped, a simple construction with easy to maintenance [2, 5]. These features, coupled with the moderate level of investment expenditure, make it a widely acknowledged solution. The simplicity of design also entails disadvantages associated with non-linear outflow of material, its uneven shape and the dimensions of the expandate. There are also instances of the die-head slots becoming clogged by large segments of the product. As a result, the device exhibits pulsations and variations in temperature of the product often approximating 20 °C. The low flexibility of the process and low susceptibility to agglomeration force a feed producer to put in place additional auxiliary devices along the processing line.

Similar characteristics are exhibited by a flange die; however, its use allows greater influence on the shape of the expandate. This type of solution can be likened to the arrangement of the die used in extruders.

As already mentioned, major work on the development of expanders is focused on the improvement of the control of baro-thermal treatment (including preconditioning), increasing the capacity to shape the expandate, increasing the quality of homogenization, and the flexibility of the processing. This has been reflected in the latest design solutions for the expander dies. A good example is a "finger-type die" of the linear outflow of the product and the so-called diafragma-disk die offering greater possibilities of shaping the expandates.

When selecting the type of expander, an important question to address is for what purposes is it needed and what kind of feed will be produced. The answer to these questions will determine the proper selection and the investment costs incurred.

11.3 Application

11.3.1 Processing

First the material is ground and preconditioned, then moves along the barrel, where it is placed under pressure inside the supporting unit and the friction created, together with the hot steam added, cause the temperature to rise. When the product is removed from the installation, it expands as a result of the rapid evaporation of the water as the pressure drops once again. The dwell time inside the expander is 5–10 s at a temperature 100–125 °C, reduced rapidly to 90 °C once outside the machine. When this process is complete, the mixture is placed inside a horizontal drier/cooler for 10 min during which the temperature is reduced to 20–24 °C in accordance with the ambient temperature.

The use of an expander in a processing line of feed pellets has many advantages, the most important being [1–3]:

- improvement of the feed nutritional quality (better feed conversion),
- more effective use of raw materials of low quality or even waste,
- sterilization of bacteriologically contaminated materials,
- better control of feeding liquid additives,
- reduced energy consumption of the pellet press,
- increasing productivity of the pellet press,
- extension of product assortment,
- possibility to use fiber materials,
- reduced emission of dust and loss during pelleting.

The disadvantage of the above-mentioned solution is the double consumption of steam and a higher energy consumption by the expander–pellet press system compared with the application of simple pelleting.

The application of an expander as an autonomous device can deliver the following results:

- greater modification of starchy ingredients,
- reduced antinutritional factors in the product,
- possible agglomeration of the product,
- improvement in the nutritional value of processed raw materials.

Figure 11.3 shows the possible applications of an expander.

The use of expandate is economically justified due to the improvement in the usability characteristics of material components applied in a wide range of feed. In order to achieve this, proper technological process parameters must be set that depend on the type of materials used and the feed application.

The results of great number of experiments can be found in the available literature concerning the influence of expansion on the physical and chemical properties of processed vegetable raw materials. Many feeding experiments *in vitro* and *in vivo* were carried out in order to check how effective was the use of expanded feed from a nutritional point of view, and how much such costly, additional baro-thermal

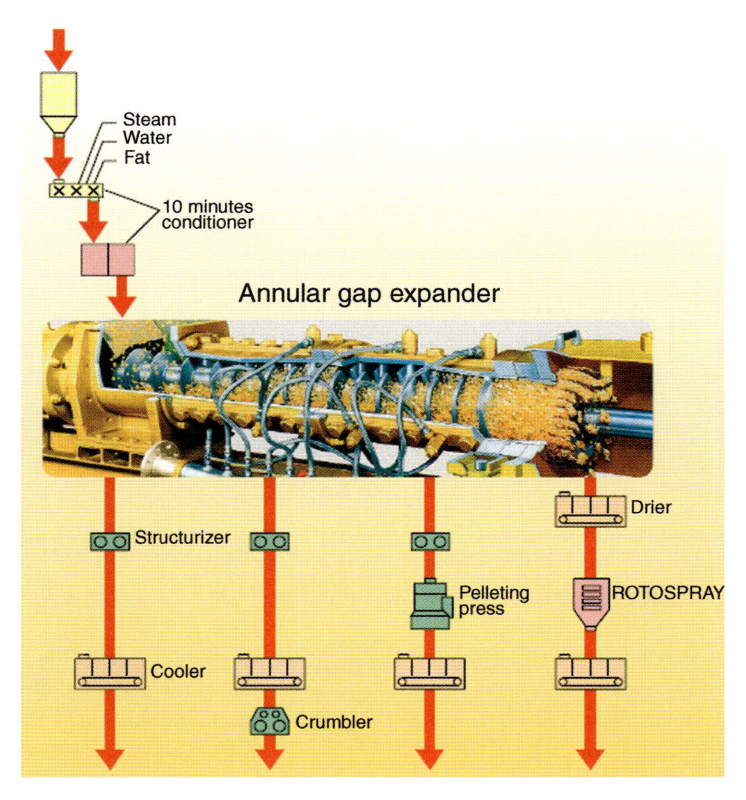

Figure 11.3 An expander with different production set-ups (permission of A. Kahl GmbH).

Table 11.2 The results of nutritional tests on poultry (3 groups of 50 broilers) [8].

Age	Expandates (115 °C)		Pellets (80 °C)	
	Weight (g)	Feed conversion factor	Weight (g)	Feed conversion factor
20 d	596	1.36	587	1.36
37 d	1620	1.75	1586	1.73
42 d	1902	1.83	1866	1.81

treatment was economically justifiable. The majority of the reports showed measured positive effects, including economic aspects but one was not questioned: improvement of feed hygienic and digestibility [6–12]. To corroborate the advantages of this technique, Table 11.2 shows the results of feeding trials on poultry fed with feed pellets pressed at approximately 80 °C and the expandate processed at 115 °C. (Unpublished results of the experiments curried at Lublin Agic. University, Poland).

It is possible to obtain quality soybeans treated using only an expander. Navarro *et al.* [11] tested the productivity yield of broiler chickens fed on diets composed of soya oil and soybean meal compared to that of broilers fed on 20% expanded beans. The final product had a KOH protein solubility level of 91.2%, 0.06 ureasic activity and a trypsin inhibitor content of 3.6 mg/g, while the values given for the 48% soya meal used as the control sample were 85.5%, 0.19 units and 2.5 mg/g respectively. Chickens fed on the expanded beans weighed 2621 g and had a conversion index of 1.938 per 2617 g while that of the control chickens was 1.988 g/g ($P > 0.05$).

The exclusive use of expansion requires a very high level of accuracy in order to ensure that the beans are processed correctly and in a uniform manner. First, they must be uniformly milled and it is advisable to use a cylinder mill instead of a hammer mill. Secondly, the beans require to be preconditioned at 100 °C for 10 min Finally, a temperature of around 130 °C must be applied for a duration of 20 s. Temperatures of 130 °C for longer times will damage the quality of the protein and, above all, the lysine availability. Data obtained during detailed studies [1] showed that proper selection of time and temperature during expansion is of great importance. It needs specific know-how to manage it and experimental trials are required.

Data obtained by Van Zuilichem *et al.* [12] on the influence of the conditions of the expansion process (preconditioning and expander temperature, among others) and the antitrypsin factor content and protein quality of the end product proved the utility of the expansion of soybeans in comparison to traditional roasting, popular in the Dutch feed sector.

In recent years, expanders have been offered a great chance resulting from their use for the enrichment of rapeseed cake – by-product of the oil press in the production of fuel bio-components. In a time of rapid development of bio-fuels, its producers have faced a problem of effective management of the vast quantities of by-products, such as the mentioned rapeseed cake. They can be used by farmers for animal feeding under one condition – their proper enrichment. In this case, one of

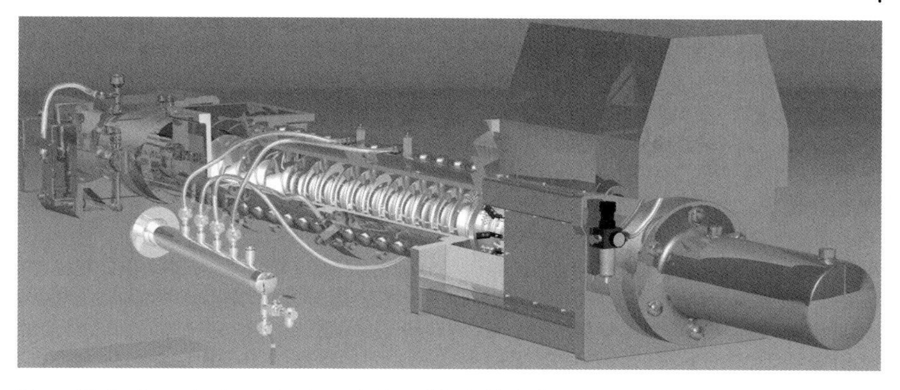

Figure 11.4 Expander/extruder (permission of A. Kahl GmbH).

the most economically advantageous solutions is a baro-thermal treatment, especially using expanders [4].

It is worth noting that expanders can also be used, to a limited extent, for the production of feed that was previously produced only by extrusion. What is meant here is pet food and aquafeed which requires an adequate treatment of mixture and final molding in the die. Due to the specific physical requirements (adequate absorption and stability in the water environment), this type of feed is produced mostly on extruders. Recently, however, expander manufacturers have begun to employ them to do the job [2]. Then, why the reservation "to a limited extent"? The reason is that expanders can only be used for simple forms of products of uncomplicated recipe and shapes. As a rule, an expander–pellet press is applied, which ultimately shapes and gives adequate physical properties to the feed. The pellets, due to the rapid absorbing ability property, are used mainly for feeding those fish species which collect food from the bottom of a reservoir. A further step in this field is the use of an expander with a replaceable die, which, once installed, works like a simple single-screw extruder. Such a solution is presented in Figure 11.4.

11.4
Concluding Remarks

The economic and nutritional aspects of the use of the expander technique make their use understandable and justifiable in the feed sector. The application of an expander–pellet press set raises the production costs by 10–15% compared with pelleting alone. Nevertheless, the overall profit and cost account is positive due to the qualitative results which counterbalance the expenses incurred.

The application of expanders in the world feed industry increases every year, which is also mirrored in the broader choice of assortment of the recently produced animal feed. The benefits of this technique are beginning to be appreciated also in the undeveloped countries. Its usefulness is fully corroborated by feed producers who have already implemented the afore-mentioned manufacturing practices.

References

1 Amandus Kahl GmbH (2000) Reduction of ANF in soybeans by means of hydrothermal treatment and expander. Amandus Kahl, Reinbek. Alemania. 8 pp.

2 Amandus Kahl GmbH (2007) Technical materials, laboratory reports (personal communication).

3 Buhler AG (2010) Technical materials of feed extrusion division (personal communication).

4 Mościcki, L., Mitrus, M., and Wojtowicz, A. (2007) *Technika ekstruzji w przetwórstwie rolno-spożywczym (in Polish)*, PWRiL, Warsaw.

5 Mościcki, L. and Stawarz, W. (1996) Ekspandery – budowa i zastosowanie (in Polish). *Zeszyty Problemowe Postępów Nauk Rolniczych*, **430**, 157–162.

6 Hancock, J.D. (2001) Extrusion technologies to produce quality pig feed. *Feed Technol.*, **5** (3), 18–20.

7 Gilbert, R. (1998) Expanding soybeans minimises ANFs. *Feed Technol.*, **2** (6), 19.

8 Mościcki, L., Kozłowska, H., Pokorny, J., and van Zuilichem, D.J. (2003) Expander cooking of rapeseed – faba bean mixtures. *EJPAU*, www.ejpau.media.pl/series/volume6/issue2/engineering/art01.html/2003.

9 Mościcki, L., Pokorny, J., Kozłowska, H., and van Zuilichem, D.J., (2004) Effect of expander cooking on SME and quality of rapeseed – faba bean mixtures. Teka Commission of Motorization Power Industry in Agriculture, vol. IV, 136–140.

10 Navarro, G.H., López, C.C., García, E., and Forat, S.M. (2001) Evaluación de la soja integral procesada mediante expansión en dietas prácticas de pollo de engorde. American Soybean Association. *Soya Noticias*, **265**, 14–22.

11 Van Zuilichem, D.J., Van der Poel, A.F.B., Cruz, U., Stolp, W., and Wolters, I. (1996) Termo mechanical treatments of soya beans. En: 2nd International Full- fat Soya Conference. ASA. Budapest, Hungary, pp. 99–117.

12
Extrusion-Cooking in Waste Management and Paper Pulp Processing

Leszek Mościcki and Agnieszka Wójtowicz

12.1
Introduction

In recent years, new methods have been developed for wet animal waste or by-products utilization management. The European Commission decisions: 2000/418/EC, 2000/766/EC, 2001/2/EC, 2001/9/EC and 2001/25/EC introduced stricter regulations on the disposal of animal waste intended for the production of animal meal; the Regulation no. 01/999 of the European Parliament introduces the terms of its use in animal nourishment [1].

Many research centers have demonstrated that the application of the extrusion-cooking technique for the utilization of waste and animal by-products allows us to use them for the high-protein components of feed or complete feed, for example, for poultry [2–7]. This can be achieved by mixing waste material with any other necessary components and producing a balanced extrudate which is able to meet the physiological needs of different species of breeding stock or pets [7, 8]. With the application of baro-thermal HTST treatment, a pasteurized product is obtained, expanded or not, which requires a relatively short drying cycle directly after leaving the extruder. In order to achieve this, some necessary production facilities are needed. A pioneer of this new technology and manufacturer of complete processing lines is Wenger Inc. from the USA [7].

The obtained product has properties that enable its storage as complete feed, or feed component, and is free from pathogenic microorganisms. A great advantage of this method is the possibility of processing a virtually unlimited range of wastes and animal by-products, which is also vital for the environment. This has been corroborated by the research carried out at the University of Life Sciences in Lublin in collaboration with the scientists from the Prague Institute of Chemical Technology [6].

Another possible application of the extrusion-cooking technique is in non-food sectors. One of the most spectacular examples can be the recent application to paper pulp production for banknote and security papers, which has revolutionized the existing methods of cellulose materials processing. Details on this subject are presented in Section 12.4.

Extrusion-Cooking Techniques: Applications, Theory and Sustainability. Edited by Leszek Moscicki

ISBN: 978-3-527-32888-8

12.2
Processing of Animal Waste

The raw materials have a significant influence on the properties of the final product. Utilization of wet post-production waste and animal by-products such as feathers, viscera, legs, heads or the whole bodies, requires special preparations prior to the extrusion-cooking process. The most important agent in relation to wastes is the material disintegration to the size of 1–2 mm in diameter. This processing aims to reduce the particle size, which facilitates the transport of mass inside the extruder. This also applies to bones added separately which should be disintegrated to a finer grinding [3, 6]. The remaining dry ingredients, such as cereals or soybean meal, should also be ground in order to facilitate water absorption and moisture equilibration. Soft waste, which can be used as a protein supplement, may be disintegrated before extrusion-cooking to a particle size of about 2 mm. After processing the obtained extrudate is crumbled after drying, which helps its application in the subsequent stages of feed production [3, 4, 7].

When selecting the processing line, it is not only maintenance costs that should be taken into account but also high throughput and versatility of the equipment. A standard manufacturing equipment should include:

- storage tanks,
- mixer,
- disintegrator,
- buffer tank,
- feeder,
- conditioner (size and function depending on the adopted technology of production),
- extruder,
- dryer,
- final grinder (additional equipment, depending on the intended use of the final product).

An extruder should have configurable screws (in the modular design) or have a set of screws with varying compression degree and geometry (in the simpler devices). The productivity of individual processing lines ranges from 1000 to 9600 kg h^{-1} for the complete feed production, and from 1200 to 11 500 kg h^{-1} for the production of feed components (moisture content of the final product at 10%) [9].

A set of equipment for the processing of animal waste into complete feed is shown in Figure 12.1.

The volume of waste material fed into the extruder depends on the moisture and fat content. Typically, the water content in these products ranges from 60 to 87% and is the main limiting factor. In the case of a complete extrusion-cooked feed, the obtained pellets must be durable and retain their properties during storage and transport. The main rule to be followed when adding animal waste or by-products is that the moisture content for the entire mixture should not exceed 25–30%. This rule should be observed during conditioning and extrusion; a common method is the addition of

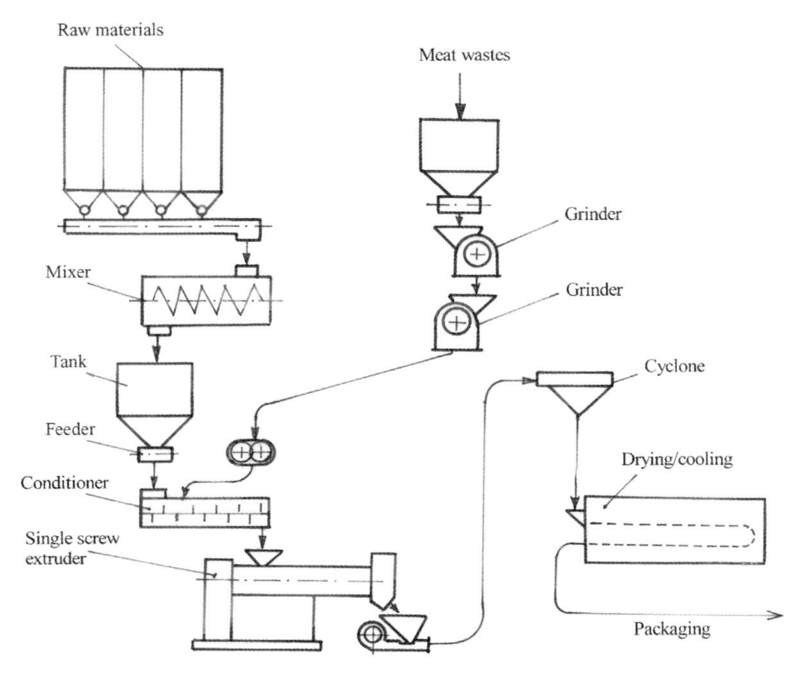

Figure 12.1 Set-up of processing line for the manufacture of feed with animal by-products [10].

steam aimed at maintaining the maximum moisture level of the processed mass at 35%. If this moisture is not kept, it is difficult to obtain a proper product shape [4, 6, 9].

Wet animal waste should be mixed with vegetable material in a proportion that will guarantee a good quality of extrudates. A typical share of by-products is 50%, with the remaining 50% being cereals, full fat soy beans meal, post-extracted soy meal and other plant materials. The addition of 18–19% of waste raw meat improves the firmness and viscosity of the obtained extrudates. Moreover, an appropriately adjusted moisture level ensures even distribution of temperatures during extrusion-cooking, which results in efficient pasteurization.

The prepared raw materials are fed into the conditioner which handles the process of moistening and thermal pre-treatment of the mixture. Next, the conditioner supplies the material to the extruder where it is cooked at a temperature from 120 to 175 °C, depending on the recipe [4, 6]. The pressure inside the extruder barrel (5–15 MPa) combined with the effect of high temperature, and average residence time 50 s, produces a standard valued extrudate. Cooking improves the taste and digestibility of the ingredients and inactivates antinutritional factors occurring particularly in the seeds of oil plants and/or leguminous plants used as filler to complement the vegetable protein components [2, 6, 9].

The majority of mixtures with post-production waste are prepared so as to maintain the moisture content at 26–35%. During extrusion, part of this moisture is lost by intensive evaporation of water after passing through the die and undergoing rapid expansion. Further reduction of the extrudate moisture occurs during drying. Spontaneous drying of the extrudate at ambient temperature is also possible but poses a risk of secondary infection of the product. Therefore, most extruded products are subjected to drying immediately after extrusion-cooking.

The primary aim of the reduced moisture level is to achieve a stable product during storage. Moreover, it is important to free the product from pathogens, which are destroyed by hot air during drying at 94–150 °C. Immediately after drying, the product should be cooled to 30–40 °C and transported to packaging or storage.

The extrusion of waste products provides an opportunity for their healthier management and resolves many intractable environmental problems. There are many advantages of extrusion that place it at the forefront of many traditional methods of utilization and feed production. The most noteworthy are:

- improved digestibility and feed efficiency for animals through the application of the HTST system, which reduces the degradation of proteins in the processed material;
- the extrusion-cooking temperature is high enough to eliminate pathogens in raw materials thus reducing feed contamination, so extrusion-cooking serves to sterilize the product;
- extrusion reduces the level of toxins;
- extrusion requires a lower energy consumption per tonne of product compared to the traditional methods of waste processing;
- extrusion requires lower initial investment compared with traditional methods of drying and waste utilization.

In EU countries, the introduction of extrusion processing for the utilization of dead animals, as is the case in the USA, is regarded as insufficient and risky for health reasons. The main criticism is the relatively short time of thermal treatment during extrusion-cooking, even including the drying phase. The EU regulations have tighter restrictions and conditions in this regard. In our view, this is justified, especially in the light of the distressing European experiences with the bovine spongiform encephalopathy disease in recent years. The problems surfacing at that time were caused by, among others, the feeding of breeding stock with animal meal. Thus, it seems well-founded that the implementation of the extrusion-cooking technique for the utilization of animal waste should be as an auxiliary and effective field of processing combined with conventional utilization methods used across Europe up till now. In this case, extrusion-cooking should be used as a constituent of a more comprehensive technology.

12.3
Utilization of Non-Meat Waste of the Food Industry

The extrusion-cooking technique can also be used to manage other waste from the food industry besides animal waste. Owing to the versatility of the extrusion machinery, its scope of application is virtually unlimited. The results of many studies show the usefulness of the extrusion technique in the processing of biomass, yeast factory wastes, distillery and brewery wastes, management of fruit pomaces and of dairy and bakery waste [15]. Extrusion provides many functional components which may be added to food and feed [10]. Counter-rotating twin-screw extruders can be effectively used to manage cellulose wastes from the paper industry and also in bran processing as a source of fiber in food and a ballast ingredient of feed. A twin-screw counter-rotating extruder, called Valeurex, developed under the Eureka program, can, owing to its universal construction, be successfully employed for the processing of fibrous materials which is impossible to carry out in standard devices [11].

There are widely known advantages of applying extrusion to manage oilseed cake. The largest application in this area may be the enrichment of oil cake, for example, extracted rape and soybean meal. Thanks to the thermal and pressure processes, the antinutritional factors in the seeds of oil plants are inactivated and the level of bitter substance in the oil is diminished. This process can be used for all oil plants (sunflower, flax, cotton, peanut) by properly adjusting the range and duration of the temperature and pressure effect in order to achieve the desired organoleptic characteristics. An extrusion-cooker can also be used as a press for crude oil, without the use of solvent extraction, which helps retain the original quantity of unsaturated fatty acids, tocopherol and lecithin.

12.4 Extrusion in Paper Pulp Processing

As was mentioned in the Section 3.3.3 the extrusion-cooking technique can be useful in the processing of cellulose and fiber-rich materials, either to decompose their chemical structure and/or to improve their application as food and feed components [12]. In recent years another interesting extrusion application has been introduced into the market – the production of paper pulp from cotton fibers, suitable for the manufacturing of banknotes and security papers [13, 14]. This extraordinary continuous pulping process BIVIS®, developed by Clextral jointly with research institutes and paper companies has already been applied successfully in France, England, Russia, China and Spain for the production of local currency banknotes.

A processing extruder consists of two identical co-rotating intermeshing self-wiping screw profiles operating within a closed barrel (Figure 12.2). Fiber separation and fiber cutting are achieved by compression and shearing forces thanks to the reverse threads screw components called reverse sections. This is a completely different method compared to the conventional defibering processes. Using a combination of reverse screw sections with varying geometries, the extruder efficiently processes virtually any cellulose raw material to achieve optimum fiber separation or cutting, depending on the raw materials.

Various chemical treatments may be performed: a chlorine-free bleaching treatment with chemicals and bleaching agents is easily and efficiently accomplished. The high consistency of the material, plus the combined actions of temperature and

Figure 12.2 Clextral TWS extruder designed for processing cellulose materials (permission of Clextral, Firminy).

pressure, accelerate the chemical reactions. Liquids or gases may be injected into the barrel in specific areas. The extrusion-cooking process greatly reduces the volume of chemicals required and dramatically lowers the volume of effluent to be treated. Bleaching to attain maximum brightness is accomplished without the use of chloric agents; sodium hydrosulfite and hydrogen peroxide achieve optimum results. Pulp washing operations consist of drawing out dissolved organic and mineral matters which remain after chemical treatment. The designed extrusion system enables one to perform both washing and defibering simultaneously thanks to filters adequately located along the high pressure zones. The extruder can be supplied with several washing sections depending on the pulp washing efficiency required. Owing to the high pressure developed and efficient mixing by the screws, highly efficient washing is possible, giving substantial savings in washing water and with much less effluents to be processed.

Compared to conventional processes, use of this system allows one to achieve [14]:

- 10 to 15% reduction in energy consumption for wood or annual plants fiber separation;
- 10 to 30% decrease in chemicals in chemical treatment and pulp bleaching, reduced water consumption for pulp washing and their polluting load allowing paper pulp manufacturers to choose a more ecologically attractive alternative and to reduce effluent treatment costs;
- reduced pulp processing time compared to traditional processes due to the specific working conditions of the machines;
- quick and easy maintenance allows reduction of the production line's pause time;

- reduced civil engineering costs: pulp lines are compact and thus require reduced civil engineering, allowing reduced costs of buildings and other associated infrastructure.

A wide range of raw materials can be successfully transformed into high quality pulp thanks to the BIVIS® extrusion-cooking processes:

- soft and hard wood;
- textile fibers such as cotton comber wastes, flax, hemp, abaca, sisal, jute, and so on;
- annual plants such as knave, wheat straw, bagasse;
- other cellulose materials such as oil palm tree wastes, sorghum and recycled fibers.

References

1 NATIONAL WASTE MANAGEMENT PLAN (2002) Appendix to the Resolution No. 219 of the Cabinet of 29 October (item 159).

2 Dolatowski, Z. and Juśko, S. (2001) Alternative technology of animal by-products utilization, Proceedings of 47th International Congress of Meat Science and Technology, Kraków, pp. 102–107.

3 Juśko, S. and Dolatowski, Z. (2001) Wpływ surowców i parametrów sterylizacji na właściwości fizyczne produktu roślinno-mięsnego. *Inż. Rol.*, **10**, 171–178 (in Polish).

4 Juśko, S., Walczak, I., Mościcki, L., and Hodara, K. (2001) Wpływ dodatku odpadów drobiowych na proces ekstruzji karmy dla psów. *Inż. Rol.*, **10**, 163–169 (in Polish).

5 Juśko, S., Mitrus, M., Mościcki, L., Rejak, A., and Wójtowicz, A., (2001) Wpływ geometrii układu plastyfikującego na przebieg procesu ekstruzji surowców roślinnych. *Inż. Rol.*, **2**, 123–129 (in Polish).

6 Mościcki, L. and Pokorny, J. (2001) Utilization of poultry wet by-products for animal feed aplication. *Sci. Agric. Biochem.* **32**, 235–243, 20.

7 Wenger Inc . (1997) Utilization of Wet By-Products – Process Description, in-house materials.

8 Wójtowicz, A. and Mościcki, L. (1998) Amerykańska metoda utylizacji odpadów produkcyjnych pochodzenia zwierzęcego. *Pasze Przem.*, **4**, 17–19 (in Polish).

9 Thomas, M., van Vliet, T., and van der Poel, A.F.B. (1998) Physical quality of pelleted animal feed, Contribution of feedstuff components. *Animal Feed Sci. Technol.*, **70**, 59–78.

10 Mościcki, L., Mitrus, M., and Wójtowicz, A. (2007) *Technika ekstruzji w przemyśle rolno-spożywczym* (in Polish), PWRiL, Warsaw.

11 Kosiba, E. (1991) Utylizacja surowców odpadowych. *Gosp. Mięs.*, **12**, 15–17 (in Polish).

12 Mościcki, L. and Zuilichem, D.J. (1986) Ekstrudowane trociny - pasza dla przeżuwaczy. Biuletyn Informacyjny Przemysłu Paszowego 51–56 (in Polish).

13 Combette, P. and Danos, L. (2009) Banknotes with BIVIS. *Clextrusion*, **18**, 9.

14 www.clextral.com (2010) Informative and technical materials.

15 Matyka, S., Mościcki, L., and Jaśkiewicz, T., (1999) Utilization of waste brewer's yeast biomass using the extrusion-cooking technique. Proceedings of 74th Inernational Congress on Agricultural Mechanization and Energy, Adana, Turkey, pp. 268–270.

13
Process Automation

Leszek Mościcki and Andreas Moster

13.1
Introduction

Roughly speaking, an extruder may be compared to a screw pump which pumps material under a specified pressure while at the same time mixing it, exposing it to shear forces and formation. An extruder can also be called a bioreactor which facilitates significant physical and chemical changes in the processed material.

Due to the nature of the processed raw materials, the automation of extruder control is complicated, mainly due to insufficient knowledge of many variable properties of the materials that come into play during the processing and their response to the impact of changeable conditions of treatment. Frequently, non-Newtonian flows are involved and radically changing rheological characteristics of the treated substances, depending on the existing conditions. This makes control engineers adopt far-reaching simplifications and assumptions in the developed mathematical models and in attempts to optimize processes.

The monitoring and control of the extrusion molding of plastics, which, as is commonly known, gave rise to the extrusion technique, is much better mastered and universally applied on an industrial scale. In the case of the extrusion of biopolymers, these issues are much more complex; no wonder that the implementation of fully automated production of extruded food and feed has been developing by small steps and some progress has only become visible in recent years.

13.2
Control and Automation

Extrusion moulders for plastics typically operate with complete filling of the inter-screw-flights space, which indicates the existence of a proportional relationship between the screw rotation and the final performance of extrusion of, for example, packaging film. This parameter is one of the most important in the operation of extruders and is taken into account in their automation control [1, 2]. During

Extrusion-Cooking Techniques: Applications, Theory and Sustainability. Edited by Leszek Moscicki

ISBN: 978-3-527-32888-8

extrusion-cooking, the most crucial and controllable factors are: the quantity of raw material used, the humidity, the process temperature, and, in some cases, the rotational speed of the screw [3–5], the main motor torque and the SME (specific mechanical energy in Wh kg^{-1}). Today's extrusion-cookers are largely controlled by simple circuit breakers due to the lack of appropriate sensors for measuring, for example, the quality of the products. This does not mean, however, that no efforts are made to control the extrusion process more smoothly and more thoroughly, especially in this time of widespread use of microprocessors in the control of parameters of technological processes [6, 7]. It appears that the most difficult operations in terms of automation are starting and stopping extrusion, due to the rapidly changing conditions of the baro-thermal treatment in the extruder. The description of the physical and chemical changes in the processed material in the form of a mathematical model aimed at optimizing the process presents many difficulties resulting from the nature of the raw materials (i.e., the ignorance of many of their variable features). However, having recourse to the intuitive calculations of many characteristics or reactions, experienced practitioners solve this problem by using so-called fuzzy modeling [4].

The primary task in the control of an extruder's work is to neutralize the persistent interruptions to the process and to maintain a stable production of a quality product. The task here is somewhat easier because the established parameters should not fluctuate under the conditions designed for the manufacture of a given product. These persistent interruptions can be divided into three categories [4, 7]:

- Distortion of high-frequency (related to the screw rotation) caused by, for example, insufficient compression, defective placing of a pressure sensor (too close to the flight) or poor construction of the die causing irregular outflow of extrudate.
- Mid-frequency interference (from 1 to 15 per min), for example, caused by interruptions in the feeding of raw materials and insufficient filling of the inter screw flights space; this significantly disrupts the process of consolidation, compressing and plasticization of the material and may lead to the development of steam traps in the processed mass.
- Low-frequency interference, that is, those that are less frequent than the material residence time in the extruder and are caused by the operation of heaters (off and on), a voltage drop, changes in water pressure or longer breaks or fluctuations in the operation of the material feeder.

In the extrusion molding of plastics, one of the most important control points is an extrusion pressure sensor in the head. The extrusion pressure has a decisive influence on the thickness and density of, for example, the produced film or profiles. On the other hand, the crucial factor in extrusion is the humidity of the raw material which clearly affects the pressure generated during extrusion. This very parameter is referred to in the design of automation processes in the extrusion of biopolymers [8].

Some disruption may be alleviated by the application of screws of appropriate structure, that is, equipped with appropriate elements (modules) of a specific geometry. For example, during the extrusion of crispbread, even a small fat content in the mixture produces pulsation (usually the fat content should not exceed 2%). This could be remedied by implementing a special plasticization zone in the modular construction of the screw; this solution allows the use of materials with a fat content of more than 5% in bread-making [9].

Unfortunately, it is impossible to implement a direct on-line control of the qualitative parameters of the extrudate such as: degree of expansion, porosity, texture, taste, nutritional value, and the ability to absorb water or fat. Nevertheless, in spite of all the automization, the operator and his experience give the highest product quality. There are, however, some measurable factors which may indirectly indicate many quality features of extruded products. These are: SME (specific mechanical energy consumption per kg of product), product temperature, pressure in the head influencing the molecular changes caused by shearing, residence time and treatment temperature [10]. They are usually directly correlated with the quality of processed extrudates. Some other factors are the engine load (amperage, torque) and screw rotation (regular rotation ensures steady pressure of extrusion).

The standard methods of control are based on the linear model of the basic process expressed with a Laplace operator transfer function. The control algorithms are usually proportional integral derivatives (PID) of analog or digital character, depending on the adopted system hardware. Sometimes the PID controller is expanded by a time-lag switch working outside the system or on-line.

In 1981 Ylikowski [11], using a PID algorithm, regulated the pressure of wheat flour extrusion in the extruder Clextral 45 by converting the received signal of the pressure sensor to a digital signal and referring it to the value established for comparison purposes. A signal deviating from the adjusted value was processed by the sensor and converted into an analogue signal in order to generate the signal regulating the engine of the flour feeder. To monitor and control the devices, the researcher used an integrated circuit on a microprocessor.

Harper [12] describes the possibility of using a more sophisticated control system called a "cascade" system for the feeding of flour, water and steam in correlation with the load of the main driving motor. The role of the operator is limited only to the input of appropriate data defining the interdependence between the fed components and the work of the motor.

In the following years, many researchers were engaged in issues of automatic control in the extrusion process and proposed more or less complex systems expediting the control of the extruders' operation work based on adaptation control or fuzzy algorithms [6, 10, 13, 14].

Figure 13.1 shows a simulation of the fuzzy control of an extruder.

Widman and Strecker [7] were the first to publish a more extensive study (in 1988) about the application of a personal computer to control the work of a twin-screw extruder by Werner & Pfleider, type ZSK 70. They built a control system handling the regulation in three circuits (Figure 13.2). The main regulated values were: consumption of energy (SME) in conjunction with water feeding, the pressure of dough in the die, and temperature (controlled by feeding steam into the barrel). When developing the process of automatic production of crispbread, they assumed that the main sources of interruption of the production process are the dispersal of the physical

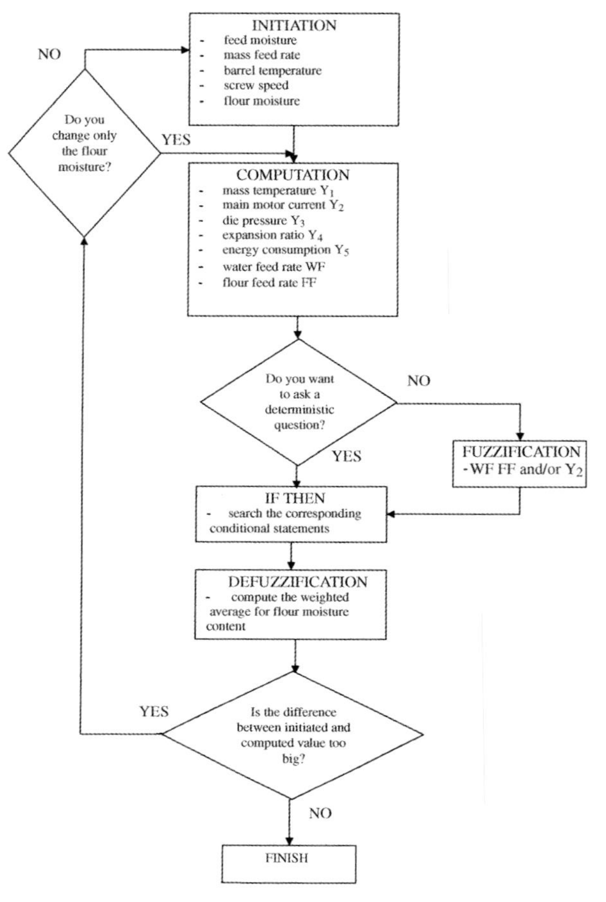

Figure 13.1 An extruder control scheme with fuzzy simulation [6, 10, 16].

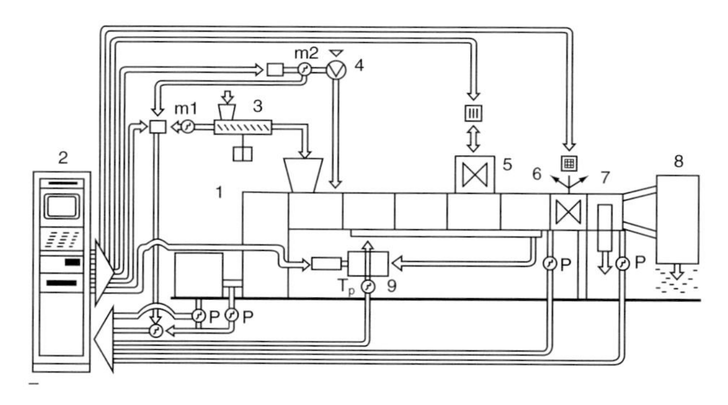

Figure 13.2 Integrated circuits for the control of extrusion of biopolymers: 1 – extruder, 2 – PC, 3 – feeding of material, 4 – feeding of water, 5 – steam valve, 6 – head, 7 – starting valve, 8 – knife, 9 – radiator for cylinder cooling fluid [7].

characteristics of raw materials and the wear (attrition) of screws and barrel in the extruder. During measurements, they concluded that, in the case of worn-out screws, only the work of three of the mentioned systems permit maintenance of acceptable quality parameters of the product. The change in flour humidity could be controlled by means of one "energy" system, yet the variations in fat content required the involvement of the "temperature" and "pressure" systems. The fluctuations in the extruder's productivity necessitated the intervention of three control systems in order to stabilize the process.

A dynamic development of computer technology in recent years has helped automate most of the technological processes that occur during the production of extruded feed and food. The basic processes include: grinding, weighing, mixing, conditioning and extrusion of mixtures of raw materials, drying and finishing (spray of flavors or coating) of extrudates.

As mentioned above, the most challenging task proved to be the correct programming of production start-up and finish. If an extruder starts too sharply and too quickly forces the prepared dose of material at fast screw rotation, the cold mass enters the compression unit of the cylinder and consolidates without plasticizing. This usually leads to a total blockage of the device, the permissible torque is exceeded (Figure 13.3). On the other hand, too slow a start-up of an extruder results in

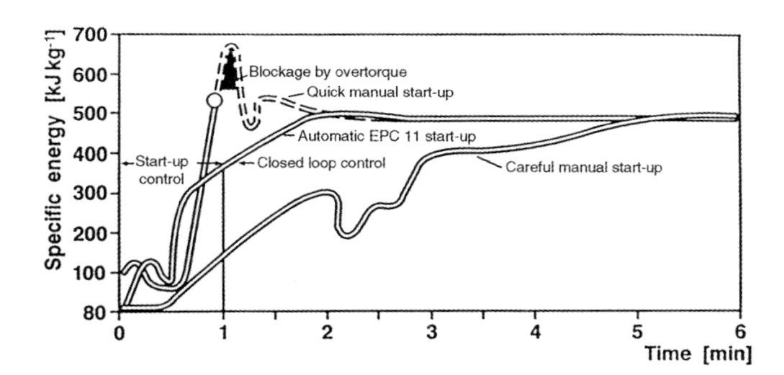

Start- up procedures: temperature curves

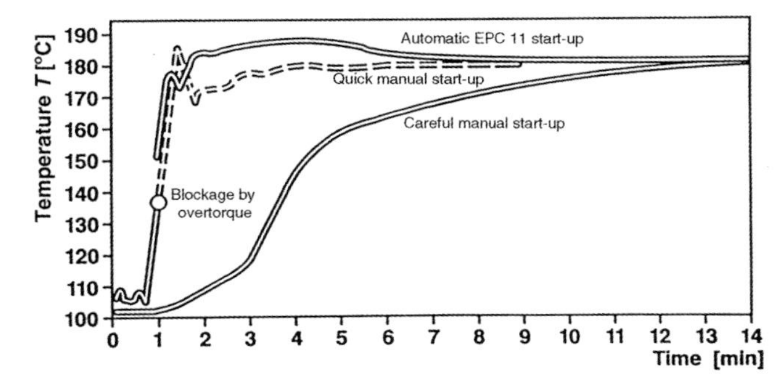

Figure 13.3 A record of energy and temperature curves during the start-up of an extruder [7].

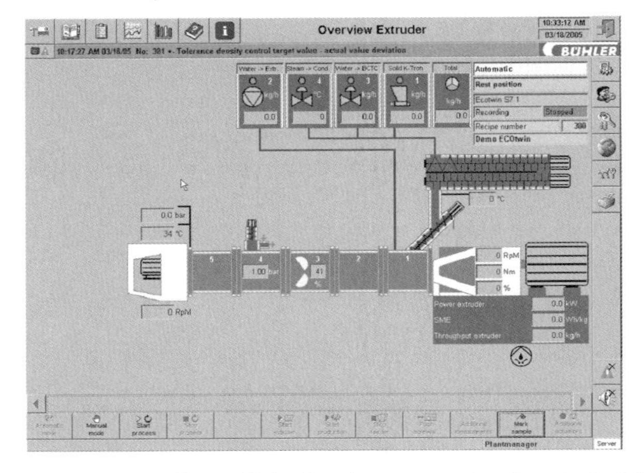

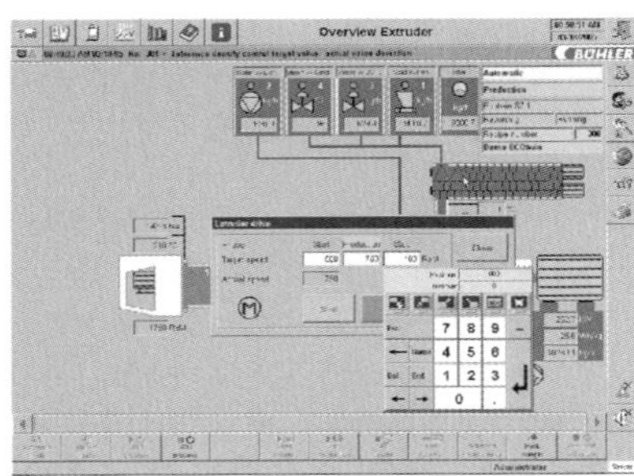

Figure 13.4 Overview to operating the extruder/adjusting of new set points on the touch screen.

unnecessary loss of material and time. In such situations, a computer with appropriate controlling software has proved indispensable.

Currently, most producers of processing lines for the production of basic extruded food and feed offer an on-demand comprehensive computer control of the manufacturing process, which costs approximately 10–15% of the value of the equipment. Referring to the example of a Swiss company, Bühler AG, we would like to introduce a control system for manufacturing extruded products. The system is based on the Siemens or Allen Bradley platform by a Touch-IPC with the Microsoft Windows operating system coupled to a PLC (programmable logic controller) system for the logic. The process control with its visualization is characterized by a very high degree of flexibility, the complete automation of the extruder process by easy operation in a Multilanguage version. Using the extruder control, an operator is able to continuously monitor the production and particular technological operations and all process values will be stored in a file for a statistical or direct showing of actual or historical trends. The system can be coupled by an Ethernet connection for copying file, data, or for print-outs. The PLC software used is intended to control and supervise the process of extrusion, sequential start and stop of production owing to the rapid transmission of digital data to PID controllers covering all the process variables (see Figures 13.4–13.6).

The extruder control enables control and supervision over the following operations with its special features:

- automatic operation mode: start and stop of production and cleaning of the machinery,
- online setting of production parameters in order to obtain high-quality product,
- regulation of the speed and torque of motors equipped with frequency inverters,
- gravimetric control of the feeding by loss and weight feeders,

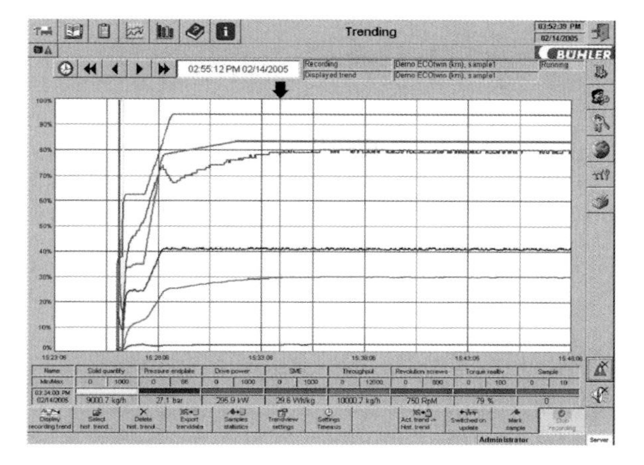

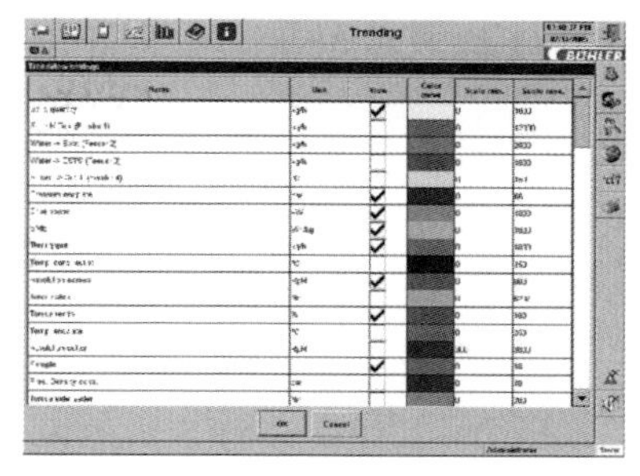

Figure 13.5 Trending and its configuration.

- manual control of the driving motors and controlling devices for servicing and maintenance,
- PID regulation of liquids or steam additions in correlation with the fed loose material,
- PID regulation of temperature zones of the barrel,
- monitoring of main motor torque with correction to prevent shut-down,
- alerting the operator to system errors, reached limitations or other warnings,
- identification of the locations and causes of failures or improper operation of assemblies or sub-assemblies,
- graphic interpretation of the relationship of variables by actual or historical trends,
- recording of all process values and alarms,
- adequate safeguards against improper or unauthorized use of the system, different access levels can be defined,
- visualization of the processing line including the monitoring of all devices,
- the option of using for the control of a processing line of high throughput in factories,
- the use of the user's language,
- recipe management before or during the process as well as parameter management for the configuration,
- production statistics with a file export,
- online help.

The system has made the grade in a number of facilities and industrial plants around the world [15]. It assists the control of the extruder's performance by adjusting the temperature of the process, appropriately feeding components and monitoring all extrusion operations determining the quality of the finished product. In other words, it enables the optimization of the process, secures the efficiency of production and controls the machinery workload and safety of the personnel.

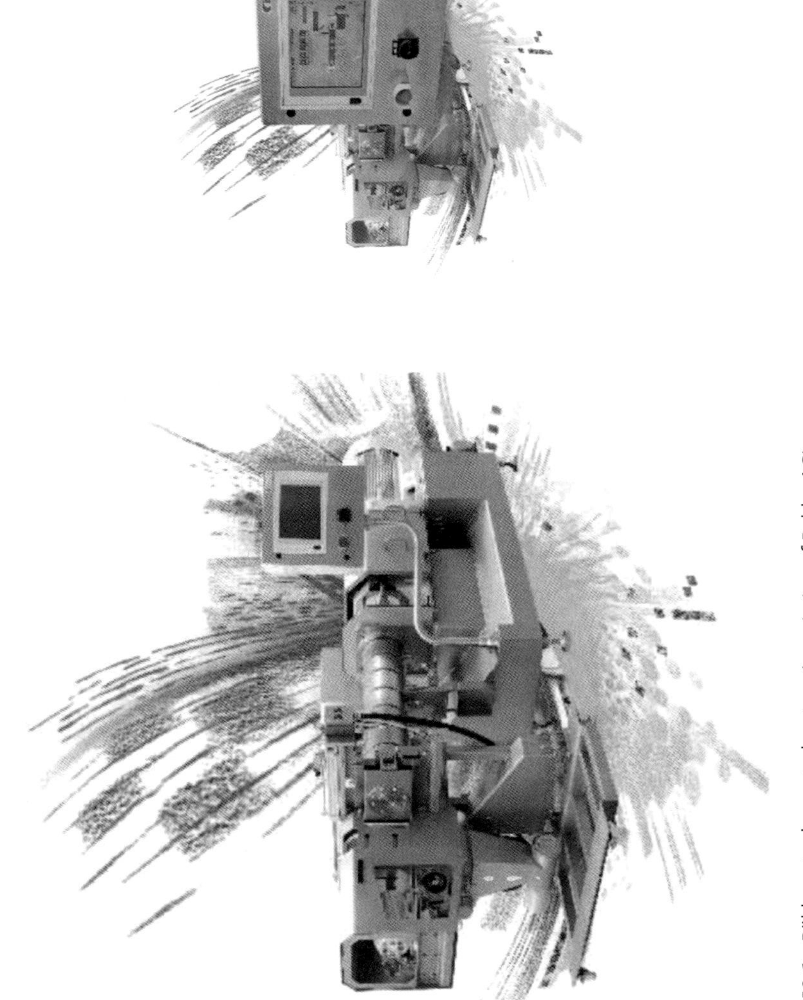

Figure 13.6 Bühler extruder control system (permission of Buhler AG).

References

1. Yang, B. and Lee, L.J. (1986) Process control of polymer extrusion, Part I: Feed-back control. *Polym. Eng. Sci.*, **26**, 197–204.
2. Yang, B. and Lee, L.J. (1986) Process control of polymer extrusion, Part II: Feedforward control. *Polym. Eng. Sci.*, **26**, 205–213.
3. Bhattachyra, M. and Hanna, M.A. (1987) Influence of process and product variables on extrusion energy and pressure requirements. *J. Food Eng.*, **6**, 153–163.
4. Eerikäinen, T., Linko, S., and Linko, P. (1988) The potential of fuzzy logic in optmization and control, in *Automatic Control and Optimization of Food Processes*, Elsevier Applied Science Publ., London, pp. 183–200.
5. Eerikäinen, T. and Linko, P. (1987) *Extrusion-Cooking Modeling, Control, and Optimalization, Extrusion Cooking*, AACC, St Paul, Minesota.
6. Mościcki, L. (2002) Automatyzacja procesów produkcyjnych ekstrudowanej żywności i pasz (in Polish). *Inż. Roln.*, **2** (35), 223–235.
7. Wiedmann, W. and Strecker, J. (1988) Process control of cooker-extruder, in *Automatic Control and Optimization of Food Processes*, Elsevier Applied Science Publ., London, pp. 201–214.
8. Levine, L., Symes, S., and Weimer, S. (1986) Automatic control of moisture in food extruders. *J. Food Eng.*, **8**, 97–115.
9. Millauer, C., Wiedmann, W.M., and Strobel, E. (1984) Extrusion cooking of dairy enriched products and modification of dairy proteins, in *Thermal Processing and Quality of Foods*, Elsevier Applied Science Publ., London, pp. 137–144.
10. Olkku, J., Hassinen, H., Antila, J., Pohjanpalo, H., and Linko, P. (1980) Automation of HTST – extrusion cooker, in *Food Process Engineering, Vol. 1, Food Processing Systems*, Applied Science Publ., London, pp. 777–790.
11. Ylikowski, K. (1981) Process control of extrusion cooker in the food manufacturing industry. M.Sc. thesis, Helsinki Univ. of Technology.
12. Harper, J.M. (1981) *Extrusion of Foods*, vol. 1, CRC Press, Boca Raton, FL.
13. Eerikäinen, T. and Linko, P. (1987) New approaches in extrusion cooking control. *Kemia-Kemi*, **14** (10b), 1051.
14. Linko, P. (1988) Uncertainties, fuzzy reasoning, and expert systems in bioengineering. *Ann. N. Y. Acad. Sci.*, **542**, 404–417.
15. www.buhlergroup.com (2010) Advertising materials and reports Bühler AG.
16. Antila, J., Pipatti, R., and Linko, P. (1984) Process control and automation in extrusion cooking, in *Thermal Processing and Quality of Food*, Elsevier Applied Science Publ., London, pp. 44–48.

14
Thermoplastic Starch

Marcin Mitrus and Leszek Mościcki

14.1
Introduction

The interest in starch as a prospective material to be used in the packaging industry grew rapidly in the 1970s as a response to the increasing public concern about the environment. It is not without significance that it was also an opportunity to reduce the production cost of plastics due to the relatively low price of starch.

For the first time, starch was used as a filler (filling additive) in plastics by Griffin [1, 2]. Polyethylene plastic films containing starch and other products based on this technology can be found on the contemporary market. Although this technology uses whole granules of starch, its participation as an additive is limited to a maximum of 10% of the mass. Starch is dried to a humidity content of less than 1% in order to prevent steam production in the extrusion process, and starch granules are surface-treated (by, for example, silicon hydrides) in order to increase the homogeneity of hydrophilic starch with the hydrophobic matrix of the plastic blend.

Almost at the same time as Griffin, Otey *et al.* [2–5] investigated starch–plastic systems, in which the starch granule structure was completely disrupted. For example, a process was developed to extrusion-blow films from mixtures of starch and poly(ethylene-co-acrylic acid) (EAA) with the addition of urea. In these materials, starch may create a continuous phase instead of being present as a molecular filler. In these composites, the starch content may go up to 50% of the mass while maintaining the desired parameters. This type of composite is manufactured and marketed by the Italian group Ferruzzi (Novamont).

The composites of starch and thermoplastic polymers can be obtained by the method of graft polymerization. Chemical treatment of starch leads to the creation of free radicals in the chain of starch that could operate in the presence of different monomers (styrene, ethylene, vinyl) as macroinitiators leading to the emergence of graft polymers of high molecular weight [6, 7].

Unfortunately, the materials obtained by these methods are not completely biodegradable. They are only subjected, after starch biodegradation, to strong

Extrusion-Cooking Techniques: Applications, Theory and Sustainability. Edited by Leszek Moscicki

ISBN: 978-3-527-32888-8

fragmentation into small pieces, which sometimes leads to the impression that the material is gradually diminishing.

Because starch is biodegradable to CO_2 and water in a relatively short time compared with most plastics, and because of the imperfectness of some existing techniques for the production of biodegradable materials, recent years have seen large-scale studies aimed at increasing the starch content in starch–plastic composites to the highest possible level. The ultimate aim of these studies is to obtain commercial, disposable items from pure starch and to exclude plastics from the formula. The ideal solution seems to be the so-called thermoplastic starch (TPS), which can be processed by traditional techniques applied in the plastics processing, namely extrusion and injection molding [5, 8–10].

In order to obtain TPS it is necessary to destroy the semi-crystalline nature of starch granules by thermal and mechanical processing. Since the melting temperature for pure dry starch is much higher than the temperature of its thermal degradation during processing, a plasticizer such as water needs to be added. Under the influence of temperature and shear forces, the natural crystalline structure of starch granules is broken and polysaccharides develop a continuous polymer phase [10–20].

14.2 Raw Materials

14.2.1 Starch

Regardless of its botanical origin, starch is a polymer of the six-carbon sugar D-glucose. The structure of D-glucose monosaccharide can be depicted in the form of both an open chain or a ring form. The ring configuration is attributed to pyranose, that is, D-glucopyranose. The pyranose ring is a more thermodynamically stable structure and is a configuration of the sugar in solution [20–23]. More information on this subject can be found in Chapter 3.

Polymerization of glucose in starch results in two types of polymers: amylose and amylopectin. Amylose is an essentially linear polymer, while the amylopectin molecule is much larger and is branched. The differences in the structure of these polymers cause significant differences in the properties and functions of starch (see Table 14.1).

14.2.2 Plasticizers

The primary role of plasticizers in the process of obtaining TPS is to facilitate the destruction of the crystalline structure of starch granules. This is possible through the lowering of the melting temperature of starch to a level lower than the temperature of its decomposition.

Table 14.1 Characteristics of amylose and amylopectin [22].

Feature	Amylose	Amylopectin
shape	linear	branched
Bond	α-1,4 (some α-1,6)	α-1,4 and α-1,6
Molecular weight	usually $< 0.5 \times 10^6$	$10 \times 10^6 - 500 \times 10^6$
Film	strong	weak
Formation of gel	strong, flexible	not gelating or weak
Color with iodine	blue	red–brown

Starch is a natural polymer containing in its molecules many hydrogen bonds between hydroxyl radicals. For this reason, it exhibits considerable tensile strength. The plasticizer acts as a diluent by increasing the distance between starch molecules and thus reducing the interaction between these molecules. As a result, the tensile strength of TPS decreases. However, the introduction of plasticizer increases the mobility of starch macromolecules, leading to elevated elasticity and elongation of derived materials.

The most common and most easily accessible plasticizer is water. Unfortunately, TPS obtained from starch plasticized only by means of water becomes very brittle and inelastic at room temperature. An additional problem associated with implementing this plasticizer is the propensity of the obtained material for expansion. This leads to the emergence of steam bubbles and empty spaces that would lower the mechanical strength of the TPS.

For these reasons, in order to improve the mechanical properties of TPS, and especially to increase its elasticity, other plasticizers are used, such as glycerol, propylene glycol, glucose, sorbitol and other [11, 15, 18, 20].

Studies have shown that in order to obtain TPS by the extrusion method, the quantity of plasticizers in the processed compounds should be from 15 to 35% of the dry weight of starch. A smaller amount of plasticizer makes the material brittle and inelastic. A larger quantity creates an elastic material but with low mechanical strength.

14.2.3 Auxiliary Substances

During the production of TPS, besides plasticizers, other additives that influence the quality of the finished material can be used. This is mainly meant to increase the mechanical strength, elasticity and color of the obtained materials. A common feature of these additives is biodegradation or neutral impact on the environment.

In order to increase the mechanical strength, the following ingredients can be added to the processed mixture: cellulose, plant fibers (flax, hemp, etc.), kaolin and bark. To improve the flexibility of the obtained materials, it is recommended to added emulsifiers such as mono- and di-glycerides of fatty acids. In order to obtain TPS of different color, the following can be added: plant fibers, kaolin and thermostable pigments [12, 13, 24–26].

14.3
Physical and Utility Features

14.3.1
Crystallographic Structure

Grain starches are made up of both amorphous and ordered areas, and the ordered areas are formed by short chains within the amylopectin molecules being arranged in clusters. Depending on the type of arrangement of the crystallographic grid in the ordered areas, there are two types of crystalline structure or a polymorphic variety that can be distinguished, namely A and B [27].

The crystallographic structure of starch depends on its plant origin. There are also several types of so-called V-type crystalline structure formed by amylose. Such crystalline systems of amylose arise after the addition to the solution of amylose of a complexing agent that facilitates the formation of the whole structure.

Depending on the process conditions and the amount of plasticizer (water, glycerol), a baro-thermal treatment is applied to produce TPS. Glycerol penetrates into the starch granules and breaks the initial crystallographic structure, which under the influence of temperature and shear forces, melts into a continuous amorphous structure. For such materials, diffraction patterns do not show the presence of a crystallographic grid. If the total thermal and mechanical energy delivered to the processed starch is not sufficient, then the obtained product reveals an unmelted residue of starch granules of clear crystallographic structure type A or B of the characteristic patterns observed on the X-ray diffraction patterns. Also, insufficient plasticizer can result in incomplete destruction of the crystallographic structure of the starch [20, 28, 29].

Immediately after processing (e.g., by the extrusion technique), TPS shows no, or only a small percentage, residue of the crystalline grid. However, several hours after production, it is possible to observe three different types of crystalline structure. These structures are generated by different arrangement of the single helices of amylase in the crystal lattice, known in the literature as the types: V_A (non-hydrated), V_H (hydrated) and E_H. The E_H-type structure is not stable and, during storage, under the influence of ambient humidity, changes into the V_H-type structure; however, the total amount of amylose crystals is not altered [20].

The quantity of emerging crystalline structures of type V depends on the parameters of the process. The extension of the material residence time leads to the rise of a greater number of crystalline structures of E_H-type, probably through a greater destruction of the structure of the starch granules and release of amylose. Similarly, the acceleration of screw rotations and the increase in the process temperature multiplies the amount of V-type crystals. Below 180 °C, V_H-type structures arise, while E_H-type structures appear above this temperature. Moreover, the content of the processed mixture has a bearing on the volume of the emerging amylose crystals. The number of amylose crystals is proportional to the amount of amylose in the starch. No amylose crystalline structures are observed in thermoplastic materials containing only amylopectin starch. V_H-type structures emerge in extruded thermoplastic starches containing more than 10% of water. E_H- and V_A-types of structures are formed in materials containing relatively little water (less than 10%). With increase in the share of plasticizer (such as glycerol) in a mixture, the volume of emerging E_H-type crystals diminishes [20].

During the storage of starch plasticized with glycerol, amylose and amylopectin recrystallize in structures of type A and B. The rate of crystallization depends on the source of the starch, the amylose content and the storage conditions. With an increase in ambient humidity, the amount of emerging crystal structures grows. Also, with increase in storage time (especially in a high humidity environment – over 60%), the process of the formation of crystals of type B becomes more intense [20, 30].

The amount of plasticizer has a significant impact on the rate of TPS crystallization during storage. A large quantity of plasticizer results in an increase in the mobility of starch chains and lowers the glass transition temperature. The rate of crystallization increases with increase in the water content. Glycerol, on the other hand, reduces the rate of crystallization with fixed water content through an interaction between starch and glycerol and by reducing both the mobility of starch chains and water stability. However, due to the hygroscopicity of glycerol, the water content usually increases and, at the same time, lowers the glass transition temperature, stimulating the rate of crystallization [3, 30].

14.3.2 Glass Transition Temperature

One of the main problems associated with starch materials is their brittleness. This defect is linked with a relatively high glass transition temperature (T_g). Glass transition temperature is the temperature of transition from a highly elastic state into the glassy state. T_g is the most important factor in determining the mechanical properties of amorphous polymers and the control of their crystallization [21, 31].

The plasticization of starch with water has been studied very frequently, also by comparison of different techniques for measuring the glass transition temperature. The most commonly used method of DSC (differential scanning calorimetry) showed T_g higher by 10–30 °C than measured by NMR (nuclear magnetic resonance) or DMTA (dynamic mechanical-thermal analysis). The analysis of the impact of water on the T_g of amylose and amylopectin showed that very branched amylopectin has a slightly lower glass transition temperature than amylose. Based on published research and practical observation, it can be concluded that starch materials containing water are mostly in the glassy state and are brittle in natural conditions [21, 32–34].

Unfortunately, the results of the measurements published by researchers often differ significantly because of complex changes occurring in starch under the influence of high temperatures and different conditions of measurement [32, 34–36].

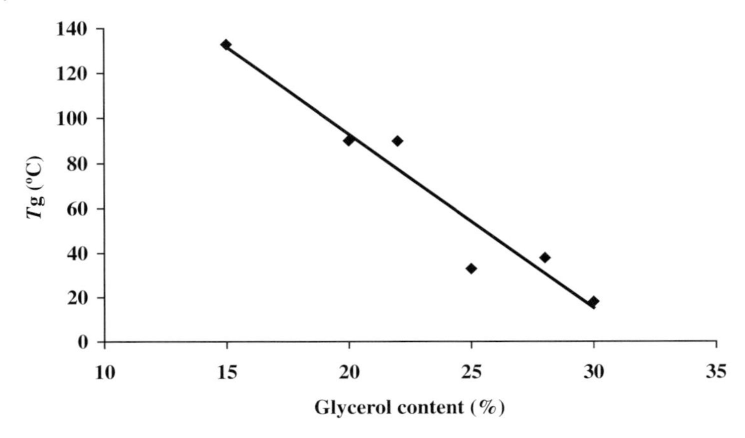

Figure 14.1 Changes in glass transition temperature of thermoplastic starch occurring with different glycerol content [42].

The glass transition temperature of TPS decreases with increase in the content of other plasticizers than water (glycerol, sorbitol). Yu *et al.* [7] report that thermoplastic maize starch with a moisture content of 10% and glycerol content of 25–35% has T_g in the range 71–83 °C. Van Soest [20] showed that thermoplastic potato starch with a moisture content of 11% and glycerol content of 26% has $T_g = 40\,°C$, and for materials with higher moisture and glycerol content T_g drops below 20 °C. In another study it has been reported that the potato starch with a moisture content of 13% and glycerol content of 15% has T_g around 25 °C, and, moreover, a mixture containing 25% of glycerol has T_g reduced to about 0 °C [7, 20, 37–40].

Our own research [41, 42] conducted with the DSC technique showed that, in the case of thermoplastic starch obtained from potato starch, the glass transition temperature decreased with increase in the glycerol content of the processed mixture.

The highest glass transition temperature, about 130 °C, was measured for mixtures with 15% of glycerol, while the lowest (about 18 °C) was for mixtures with 30% glycerol (Figure 14.1). The obtained results are principally in line with the test results obtained by other researchers [30, 39, 43].

14.3.3 Mechanical Properties

The mechanical properties of TPS depend on the production temperature, the quantity of water contained and the quantity and type of added plasticizers and auxiliary substances. The greatest impact on changes in mechanical properties is attributed to the amount of plasticizer and auxiliary substance used.

The most commonly used plasticisers, such as glycerol, glycol and sorbitol, have the same hydroxyl radicals as starch, and thus they are compatible with starch macromolecules. With increasing plasticizer content, the tensile strength of TPS decreases, and, at the same time, its elongation grows. However, this occurs only up to a certain percentage of glycerol. If the glycerol content in a mixture exceeds 35%, a decrease in the elongation of TPS is visible. This is because at too high a percentage of glycerol the intermolecular interactions are very weak because the interactions between starch molecules are replaced by interactions between glycerol and starch molecules [21, 37, 38, 40, 44, 45].

With increasing water content in the mixture, the tensile strength of thermoplastic starch and the elongation improves. If, however, the percentage of water exceeds 35%, the mechanical strength of TPS is reduced [14, 45].

With increase in the content of auxiliary substances such as cellulose fibers, linen, kaolin and pectin, the mechanical strength is enhanced but the TPS elongation is reduced. The addition of urea or boric acid increases the elongation, at the same time reducing the mechanical strength of TPS [7, 13, 40, 46].

During the storage of products made of thermoplastic starch, the phenomenon of recrystallization of amylose and amylopectin occurs. With longer storage time, and thus the crystallinity of TPS, the mechanical strength of the material is enhanced and the elongation is reduced. The increase in the surrounding humidity during storage has a clearly adverse effect on the mechanical properties of starch material [20].

14.3.4
Rheological and Viscoelastic Properties

Regardless of the type and amount of plasticizer used, melted thermoplastic starch exhibits the non-Newtonian flow characteristic of pseudoplastic fluids. With increasing content of plasticizer, the viscosity decreases and gives way to the growing flow ability of TPS materials. Thermoplastic starch shows exponential dependence of viscosity on temperature, which is well known in the case of conventional polymers [7, 11, 26, 29, 40, 44].

The most common method of testing the viscoelastic properties of materials is DMTA analysis. This method explores the changes in storage modulus (E') and loss modulus (E'') and the internal friction coefficient ($\tan\delta = E''/E'$) as a function of temperature for various frequencies of mechanical stress. E' is a measure of the flexibility and elasticity of material, the higher the component, the more flexible the material. E'' is a measure of energy dissipated in the material in the form of heat. This technique is one of the most important methods of determining the glass transition temperature of plastics and TPS.

14.3.5
Water Absorption

The research on water absorption by starch without the participation of plasticizer showed that amylopectin absorbs less water than amylose. The influence of the addition of glycerol on water absorption is similar with the two components of starch [32, 34].

At ambient relative humidity below 50%, the water content for both amylopectin and amylose was lower in the mixtures containing glycerol. This phenomenon is probably related to the replacement of structurally immobilized water by glycerol. Similar behavior was also observed for starch plasticized with sorbitol.

At ambient relative humidity higher than 50%, the mixtures containing the largest share of glycerol showed the highest water content. When the ambient humidity exceeded 70%, the water content in starch plasticized with glycerol was higher than in the starch without glycerol [18, 34, 39, 43]. Although both components of starch react similarly during water absorption, with high ambient humidity amylose absorbs more water than amylopectin. This is most likely due to the crystallization of amylopectin because crystallization reduces the water absorption of hydrophilic polymers.

Absorption of water by thermoplastic starch is one of the major problems during the production of packaging materials. During manufacturing and storage of TPS products, they may absorb water from the environment, or release it through desorption (evaporation), thereby changing their mechanical properties. This is a huge problem because it is difficult to ensure sustained and stable conditions for production and storage and, hence, the properties of the finished products. To minimize the impact of water absorption, some addition of hygroscopic substance is recommended which drains the processed material, or covers the surface of finished TPS products with a coating forming a good barrier to water.

14.4 Production of Biodegradable Packaging Materials

Packaging materials with TPS can be obtained by a one- or two-step method. In the one-step method, all the mixture components are fed directly into a device, such as an extruder, that produces packaging material. If the two-step method is used, in the first step, a ready-made mixture of ingredients is fed into an extruder to produce a TPS granulate, which is a semi-product for the manufacture of packaging. In step two, the granulate is processed into packaging using equipment for the processing of plastics.

The two-step method, widely used in plastics processing, seems to be a good solution for the producers of both packaging and extruded products. Packaging producers, by buying a semi-product as granulate, save production and warehouse space and do not need to have the equipment and ingredients for the manufacture of TPS granulate. On the other hand, the manufacturers of extruded products are provided with an additional assortment of items produced on their extruders. This is particularly important for small and medium-sized enterprises, especially as the extrusion of TPS involves relatively low energy inputs, amounting to an average of $0.07\,kWh\,kg^{-1}$. If observing a proper technological regime, even single-screw extruders with $L/D = 16/1$ may be suitable for production [47, 48].

14.4.1 Protective Loose-Fill Foams

Generally, the extrusion technique can be successfully employed for production of starch-based foams. The physical properties of loose-fills, such as density, porosity, cell structure, water absorption characteristics and mechanical properties are highly dependent on the raw materials and additives. The mechanical behavior of foamed pellets can be adjusted effectively by controlling the cell structure through using different additives. At room temperature and 50% relative humidity, some mechanical properties, such as compressive strength or compressive modulus of elasticity are comparable to commercial EPS foams.

Starch-based foams can be prepared from different starch sources, replacing 70% polystyrene with biopolymer starch. Functional starch-based plastic foams can be prepared from different starch sources depending on their availability.

Although the use of starch in loose-fill products gives advantages in the form of biodegradability and environmental protection, these products have been criticized for their imperfection compared with EPS loose-fill products. EPS- and starch-based foams have differences, but the differences do not compromise performance.

These products differ with respect to composition and method of manufacture. Foam and bulk densities, which are higher by a factor of two to three times than either EPS-based foams, are attributable to the density of starch, which is 50% higher than polystyrene homopolymer and to the direct water-to-steam expansion process, which creates a predominately open cellular structure that stops foam expansion. Starch-based foam loose-fill is very hygroscopic. The foam density of starch-based products is significantly increased, between 10 and 30%, after conditioning at high humidity [49, 50].

The compressive stress of most starch-based foams does not differ significantly from EPS products. Chemically modified starches give foams with good retention of compressive stress over a broad humidity range [51, 52].

The resiliency of starch-based foams with values between 69.5 and 71.2% are, as a group, about 10% lower than EPS foams. Although starch-based foams absorbs 13 to 16 wt% moisture after conditioning at 80% r.h. and 23 °C, these products retain between 62 and 67% resiliency [50].

14.4.2 Film Blowing

TPS granulate made by the extrusion technique can be successfully used as a semi-product for the manufacture of biodegradable film for agricultural and packaging applications. Such a product is obtained by extrusion-blowing on standard extrusion machines used for plastics (Figure 14.2) [53]. During production, it is necessary to observe a strict technological regime, in particular when it comes to the distribution of temperature in different sections of the extruder.

14.4.3 Production of Shaped-Form Packagings

Thermoplastic starch can be effectively used for the production of various forms of rigid packaging. Such products can be obtained through the use of an injection

Figure 14.2 Production of TPS film in the laboratory of Food Process Engineering Department, Lublin University of Life Sciences.

molding technique, a widespread method in plastics processing (Figure 14.3). Materials obtained by this method from TPS exhibit a relatively good mechanical strength. The best offer mechanical strength comparable with the durability of polystyrene products (Figure 14.4).

Figure 14.3 A standard injection molding machine.

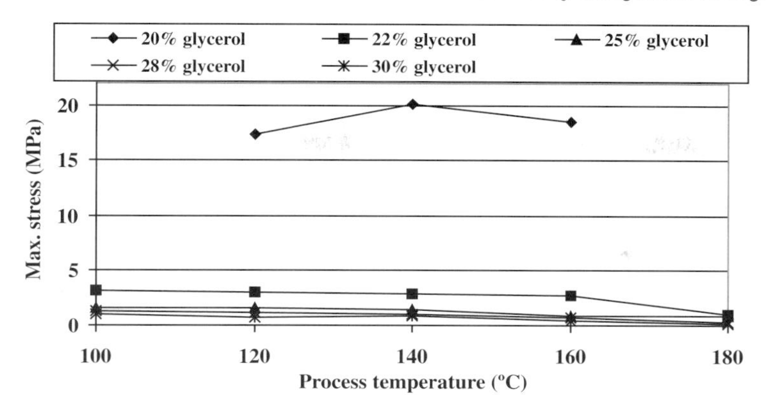

Figure 14.4 Sample results of strength tests of TPS mouldings obtained by injection molding [54].

TPS products display limited original shrinkage, hence they retain a stable shape and desired dimensions. An additional advantage is the option of supplementing the recipe with additives such as plant fiber in order to increase the mechanical strength of the products and stabilize their shape [8, 54, 55].

Biodegradability tests show that TPS moldings stored in soil for 4 weeks lost 20–30% of their mass and were partly disintegrated (Figure 14.5). After 12 weeks of storage underground, the moldings were completely decomposed [8, 54].

Because TPS products are biodegradable, the injection molding technique can be used for the production of pots and containers used for keeping cultivated plants, young trees and shrubs. Rigid forms of TPS can be used as disposable containers to store dry materials and other small articles of everyday use, such as clips, and thus helps the replacement of traditional plastics.

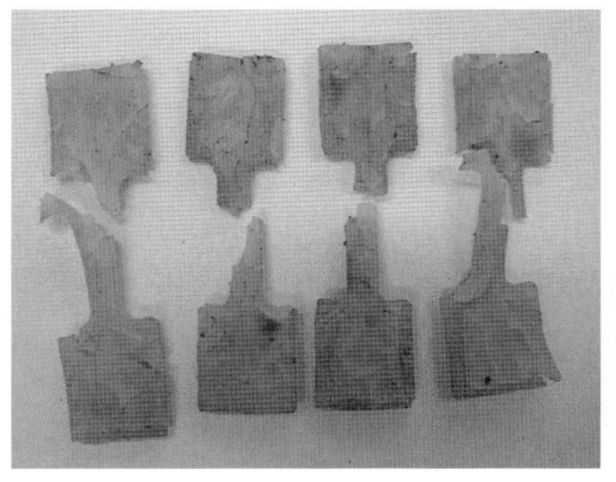

Figure 14.5 TPS moldings kept in soil for 4 weeks [55].

14.5
Concluding Remarks

Starch has been considered as a material candidate in certain thermoplastic applications because of its known biodegradability, availability and low cost. Agricultural crops provide different sources of biopolymers (starch, protein, and cellulose) which can be readily used to make biodegradable plastics. When biodegradability is required, thermoplastic starch can be an alternative material for replacement of many petroleum-based products and has gained much attention. Development of practical TPS resins includes the addition of processing aids and plasticizers to aid gelatinization during processing, thus producing suitable mechanical properties in the finished product. Unfortunately, TPS has two main disadvantages: poor mechanical properties and water resistance, which limit its application.

Concerning the production of biodegradable thermoplastic material in an extruder, the product includes a natural polymer, a plasticizer and an inorganic or organic compound such as fillers or fibers. Their presence keeps the plasticizer in the material, and hence the mechanical properties of the composite TPS are more stable than those in other biodegradable thermoplastic materials. Studies with biodegradable starch-based polymers have recently demonstrated that these materials have a range of properties, which make them suitable for use in several biomedical applications, ranging from bone plates and screws to drug delivery carriers and tissue engineering scaffolds. The primary purpose was to use TPS to produce films for agricultural and land applications. TPS is useful for the manufacture of commercial articles, either by injection molding or film blowing. TPS can be used as biodegradable foams applied extensively as cushioning materials for the protection of fragile products during transportation and handling. TPS films possess low permeability and thus become attractive materials for food packaging.

References

1 Griffin, G.J.L. (1994) *Chemistry and Technology of Biodegradable Polymers*, Chapman and Hall, Glasgow.

2 Röper, H. and Koch, H. (1990) The role of starch in biodegradable thermoplastic materials. *Starch*, **42**, 123.

3 Otey, F.H. and Mark, A.M. (1976) Degradable starch-based agricultural mulch film, US Patent 3,949,145.

4 Otey, F.H., Westhoff, R.P., and Doane, W.M. (1980) Starch-based blown films. *Ind. Eng. Chem. Prod. Res. Dev.*, **19**, 592.

5 Wiedmann, W. and Strobel, E. (1991) Compounding of thermoplastic starch with twin-screw extruders. *Starch* (43), 138.

6 Chinnaswamy, R. and Hanna, N.A. (1991) Extrusion-grafting onto starch vinylic polymers. *Starch*, **43**, 396.

7 You, X., Li, L., Gao, J., Yu, J., and Zhao, Z. (2003) Biodegradable extruded starch blends. *J. Appl. Polym. Sci.*, **88**, 627.

8 Janssen, L.P.B.M. and Mościcki, L. (eds) (2009) *Thermoplastic Starch. A Green Material for Various Industries*, Wiley-VCH Verlag GmbH., KGaA, Weinheim.

9 Mościcki, L., Janssen, L.P.B.M., and Mitrus, M. (2006) Przetwórstwo skrobi termoplastycznej na cele opakowaniowe

(in Polish). *Inżynieria Rolnicza*, **6** (81), 65–72.
10 Shogren, R.L., Fanta, G.F., and Doane, W.M. (1993) Development of starch based plastics – a reexamination of selected polymer systems in historical perspective. *Starch*, **45**, 276.
11 Aichholzer, W. and Fritz, H.G. (1998) Rheological characterization of thermoplastic starch materials. *Starch*, **50**, 77.
12 Avérous, L., Fringant, C., and Moro, L. (2001) Starch-based biodegradable materials suitable for thermoforming packaging. *Starch*, **53**, 368.
13 De Carvalho, A.J.F., Curvelo, A.A.S., and Agnelli, J.A.M. (2001) A first insight on composites of thermoplastic starch and kaolin. *Carbohyd. Polym.*, **45**, 189.
14 Hulleman, S.H.D., Janssen, F.H.P., and Feil, H. (1998) The role of water during plasticization of native starches. *Polymer*, **39**, 2043.
15 Lörcks, J. (1998) Properties and applications of compostable starch – based plastic material. *Polym. Degrad. Stabil.*, **59**, 245.
16 Mitrus, M. (2006) Microstructure of thermoplastic starch polymers. *Int. Agrophys.*, **20**, 31–35.
17 Mościcki, L. and Mitrus, M. (2004) Biodegradowalne biopolimery: materiały opakowaniowe (in Polish). *Inżynieria Rolnicza*, **5** (60), 215–222.
18 Nashed, G., Rutgers, R.P.G., and Sopade, P.A. (2003) The plasticisation effect of glycerol and water on the gelatinisation of wheat starch. *Starch*, **55**, 131.
19 Stepto, R.F.T. (1997) Thermoplastic starch and drug delivery capsules. *Polym. Int.*, **43**, 155.
20 Van Soest, J.J.G. (1996) Starch plastics: structure – property relationships, PhD thesis Utrecht University, Utrecht.
21 De Graaf, R.A., Karman, A.P., and Janssen, L.P.B.M. (2003) Material properties and glass transition temperatures of different thermoplastic starches after extrusion processing. *Starch*, **55**, 80.
22 Thomas, D. and Atwell, A. (1999) *Starches*, Eagan Press, St. Paul.
23 Zobel, H.F. (1988) Molecules to granules: A comprehensive starch review. *Starch*, **2**, 44.
24 Funke, U., Bergthaller, W., and Lindhauer, M.G. (1998) Processing and characterization of biodegradable products based on starch. *Polym. Degrad. Stabil.*, **59**, 293.
25 Ge, J., Zhong, W., Guo, Z., Li, W., and Sakai, K. (2000) Biodegradable polyurethane materials from bark and starch. I. Highly resilient foams. *J. Appl. Polym. Sci.*, **77**, 2575.
26 Yu, L., Christov, V., Christie, G., Gray, J., Dutt, U., Harvey, T., Halley, P., Coombs, S., Jayasekara, R., and Lonegan, G. (1999) Effect of additives on gelatinization, rheological properties and biodegradability of thermoplastic starch. *Macromol. Symp.*, **144**, 371.
27 Bogracheva, T.Y., Wang, Y.L., Wang, T.L., and Hedley, C.L. (2002) Structural studies of starch with different water contents. *Biopolymers*, **64**, 268.
28 Serghat – Derradji, H., Copinet, A., Bureau, G., and Couturier, Y. (1999) Aerobic Biodegradation of extruded polymer blends with native starch as major component. *Starch*, **51**, 369.
29 Yu, J., Gao, J., and Lin, T. (1996) Biodegradable thermoplastic starch. *J. Polym. Sci.*, **62**, 1491.
30 Garcia, M.A., Martino, M.N., and Zaritzky, N.E. (2000) Microstructural characterization of plasticized starch-based films. *Starch*, **4**, 118.
31 Broniewski, T., Kapko, J., Płaczek, W., and Thomalla, J. (2000) *Metody badań i Ocena Właściwości Tworzyw Sztucznych (in Polish)*, Wydawnictwa Naukowo-Techniczne, Warszawa.
32 Bizot, H., Le Bail, P., Leroux, B., Davy, J., Roger, P., and Buleon, A. (1997) Calorimetric evaluation of the glass transition in hydrated, linear and branched polyanhydroglucose compounds. *Carbohyd. Polym.*, **32**, 33.
33 Moates, G.K., Noel, T.R., Parker, R., and Ring, S.G. (2001) Dynamic mechanical and dielectric characterisation of

amylose–glycerol films. *Carbohyd. Polym.*, **44**, 247.

34 Myllärinen, P., Partanen, R., Seppälä, J., and Forssell, P. (2002) Effect of glycerol on behaviour of amylose and amylopectin films. *Carbohyd. Polym.*, **50**, 355.

35 Bindzus, W., Livings, S.J., Gloria – Hernandez, H., Fayard, G., van Lengerich, B., and Meuser, F. (2002) Glass transition of extruded wheat, corn and rice starch. *Starch*, **54**, 393.

36 Shogren, R.L. (1992) Effect of moisture content on the melting and subsequent physical aging of cornstarch. *Carbohyd. Polym.*, **19**, 83.

37 Gaudin, S., Lourdin, D., Le Botlan, D., Ileri, J.L., and Colonna, P. (1999) Plasticisation and mobility in starch–sorbitol films. *J. Cereal Sci.*, **29**, 273.

38 Lourdin, D., Bizot, H., and Colonna, P. (1997) Antiplasticization in starch–glycerol films? *J. Appl. Polym. Sci.*, **63**, 1047.

39 Lourdin, D., Coignard, L., Bizot, H., and Colonna, P. (1997) Influence of equilibrium relative humidity and plasticizer concentration on the water content and glass transition of starch materials. *Polymer*, **38**, 5401.

40 Yu, J., Chen, S., Gao, J., Zheng, H., Zhang, J., and Lin, T. (1998) A study on the properties of starch/glycerine blend. *Starch*, **50**, 246.

41 Mitrus, M. (2004) Wpływ obróbki barotermicznej na zmiany właściwości fizycznych biodegradowalnych biopolimerów skrobiowych (in Polish). PhD thesis, Akademia Rolnicza w Lublinie, Lublin.

42 Mitrus, M. (2005) Glass transition temperature of thermoplastic starch. *Int. Agrophys.*, **19**, 237–241.

43 Forssell, P.M., Mikkilä, J.M., Moates, G.K., and Parker, R. (1997) Phase and glass transition behaviour of concentrated barley starch–glycerol–water mixtures, a model for thermoplastic starch. *Carbohyd. Polym.*, **34**, 275.

44 Liu, Z.Q., Yi, X.S., and Feng, Y. (2001) Effects of glycerin and glycerol monostearate on performance of thermoplastic starch. *J. Mater. Sci.*, **36**, 1809.

45 Shogren, R.L. (1993) *Effects of Moisture and Various Plasticizers on the Mechanical Properties of Extruded Starch, in Biodegradable Polymers and Packaging* (eds Ch. Ching, D.L. Kaplan, and E.L. Thomas), Technomic Publishing Company Inc., Lancaster-Basel, USA-Switzeland.

46 Fishman, M.L., Coffin, D.R., Konstance, R.P., and Onwulata, C.I. (2000) Extrusion of pectin/starch blends plasticized with glycerol. *Carbohyd. Polym.*, **41**, 317.

47 Mitrus, M. (2005) Changes of specific mechanical energy during extrusion cooking of thermoplastic starch. *TEKA Kom. Mot. Energ. Roln.*, **5**, 152–157.

48 Mitrus, M. (2006) Investigations of thermoplastic starch extrusion cooking process stability. *TEKA Kom. Mot. Energ. Roln.*, **6**, 138–144.

49 Bhatnagar, S. and Hanna, M.A. (1995) Properties of extruded starch-based plastic foam. *Ind. Crop. Prod.*, **4**, 71–77.

50 Tatarka, P.D. and Cunningham, R.L. (1998) Properties of protective loose-fill foams. *J. Appl. Polym. Sci.*, **67**, 1157–1176.

51 Bhatnagar, S. and Hanna, M.A. (1996) Starch-based plastic foams from various starch sources. *Cereal Chem.*, **75**, 601–604.

52 Zhou, J., Song, J., and Parker, R. (2006) Structure and properties of starch-based foams prepared by microwave heating from extruded pellets. *Carbohyd. Polym.*, **63**, 466–475.

53 Rejak, A. and Mościcki, L. (2004) Właściwości fizyczne skrobiowych folii opakowaniowych (in Polish). *Inżynieria Rolnicza*, **5** (60), 299–304.

54 Oniszczuk, T. (2006) Wpływ parametrów procesu wtryskiwania na właściwości fizyczne skrobiowych materiałów opakowaniowych (in Polish). PhD thesis, Akademia Rolnicza w Lublinie, Lublin.

55 Oniszczuk, T., Janssen, L.P.B.M., and Mościcki, L. (2006) Wpływ dodatku włókien naturalnych na wybrane właściwości mechaniczne wyprasek biopolimerowych (in Polish). *Inżynieria Rolnicza*, **6** (81), 101–108.

15
Scale-Up of Extrusion-Cooking in Single-Screw Extruders

Leon P.B.M. Janssen and Leszek Mościcki

15.1
Introduction

Scale-up rules provide the possibility to transfer knowledge obtained on small scale laboratory equipment to large scale production units. The principle of scale-up is that equations describing the behavior of process equipment can be written in a dimensionless form. If the resulting dimensionless groups are kept equal in the small scale and in the large scale equipment, the solutions of the various equations remain constant in a dimensionless form. In extrusion cooking the significance of scale-up is twofold: it provides the possibility for product and process development on a small scale before doing the final trials on the production equipment and it gives the possibility to have a smooth transition to new equipment if a significant increase in production is needed.

Scale-up of extrusion-cooking suffers from the same general problems that are encountered in many other processes in the process industry:

- on scaling up, the surface to volume ratio decreases and therefore the possibilities for heat transfer are limited in large scale equipment,
- at equal temperature differences the temperature gradients, and therefore the heat fluxes, are smaller in large scale equipment,
- at equal shear fields in large scale and small scale equipment diffusion limitations connected to distributive mixing can be more predominant in large food extruders.

Various theories on the scale-up of single-screw food extruders exist. Due to the high viscosities a considerable amount of the process energy is transferred into heat by viscous dissipation. Therefore, the thermal considerations will dominate the scale-up rules and an important aspect is the extent to which the process is adiabatic or not. If the process can be considered to occur adiabatically a sufficient condition for scale-up will be that the energy input per unit throughput is constant and the average temperature of the end product will be the same in the small scale and the large scale

Extrusion-Cooking Techniques: Applications, Theory and Sustainability. Edited by Leszek Moscicki

ISBN: 978-3-527-32888-8

equipment. If this is not the case, similar temperature profiles in both types of equipment, called complete thermal similarity, are required.

The degree to which the process is adiabatic can be estimated from the Brinkmann number, that can be rewritten for extruders as [2, 12]:

$$Br = \frac{\mu v^2}{\lambda \Delta T} = \frac{\mu(\pi N D)^2}{\lambda \Delta T} \tag{15.1}$$

where λ is the thermal conductivity of the starch mass ($W\,m^{-2}\,K^{-1}$) and ΔT is the temperature difference between the mass and the barrel wall. If this Brinkmann number is much larger than unity, adiabatic scale-up is acceptable.

A particular dependence is the quadratic occurrence of the diameter. This implies that the Brinkmann number is generally large for production machines. It is generally not possible to keep the Brinkmann number constant for large scale and small scale machines. To obtain reliable predictions on small scale machines the Brinkmann number for this machine should at least be much larger than unity, which set its limitations to the minimum screw diameter of the small scale machine. If this number is smaller that unity for laboratory machines reliable scale-up is not possible.

In order to obtain complete thermal similarity, the screw rotation rate has to decrease drastically with increasing screw diameter, as compared to the adiabatic case [7, 9]. As a result, the scale factor for the throughput is only 1.5 for Newtonian fluids (and decreases even further for fluids with pseudo-plastic behavior). This scale-up factor (q) for the throughput is defined from:

$$\left[\frac{Q}{Q_0}\right] = \left[\frac{D}{D_0}\right]^q \tag{15.2}$$

where Q denotes the throughput, D the screw diameter and the subscript 0 indicates the small extruder. In the case of adiabatic scale-up a scale-up factor up to 3 can be achieved. For a standard industrial extruder series in a first approximation it may be stated that:

$$\left[\frac{Q}{Q_0}\right] = \left[\frac{D}{D_0}\right]^{2.8}$$

When an extruder is scaled it is important to keep the process in the large machine as much as possible similar to that in the small machine. Complete similarity is often not possible or is impractical and choices in similarity have to be made. Several types of similarities can play a role in the scale-up of an extruder:

- *Geometric similarity* exists if the ratio between any two length parameters in the large scale equipment is the same as the ratio between the corresponding lengths in the small scale model. This is not necessarily the case, as will be seen later, but in general this condition can be very convenient.
- *For hydrodynamic similarity* two requirements should be fulfilled: The dimensionless flow profiles should be equal and for twin screw extruders, both extruders

should have the same (dimensionless) filled length. Equal dimensionless flow profiles lead to equal shear rates in corresponding locations, but not to equal velocities.

- *Similarity in residence times* means equal residence times in the small scale and large scale equipment. This is a requirement that is often not fulfilled in extrusion processes and in thermoplastic starch extrusion this can only be realized if the scale-up is adiabatic.
- Absolute *thermal similarity* is difficult to achieve, as stated before. This similarity indicates equal temperatures in all corresponding locations. A distinction has to be made between processes with small heat effects and those with high heat effects. For adiabatic processes where the heat generation is far more important than heat removal to the wall, similarity based on over-all energy balances is generally used. Although, strictly speaking, this does not lead to thermal similarity, equal average end temperatures of the product lead to far more favorable scale-up rules.

15.2 Basic Analysis

To derive rules for scale-up, all parameters are assumed to be related to the diameter ratio by a power-relation. For this purpose, in this chapter all basic parameters will be written in capitals, whereas the scale-up factors will be written in small print. This implies that all relevant parameters can be related to the screw diameter according to:

$$\left[\frac{N}{N_0}\right]=\left[\frac{D}{D_0}\right]^n;\left[\frac{P}{P_0}\right]=\left[\frac{D}{D_0}\right]^p;\left[\frac{\mu}{\mu_0}\right]=\left[\frac{D}{D_0}\right]^v;$$
$$\left[\frac{H}{H_0}\right]=\left[\frac{D}{D_0}\right]^h;\left[\frac{L}{L_0}\right]=\left[\frac{D}{D_0}\right]^l;\left[\frac{\tau}{\tau_0}\right]=\left[\frac{D}{D_0}\right]^t;\left[\frac{R}{R_0}\right]=\left[\frac{D}{D_0}\right]^r \tag{15.3}$$

where N is the rotation rate of the screws, P the die pressure, and μ the viscosity. The back flow Q_b is an important parameter in scaling up. For closely intermeshing twin screw extruders it signifies the total amount of leakage flows, for single screw, self-wiping and non-intermeshing extruders it signifies the pressure flow. H denotes the channel depth, L the screw length, τ the residence time in the extruder and R the pumping efficiency [1, 3, 10].

For thermal scale up rules two more parameters have to be used, the Greaz number (Gz) that will be defined later and the Brinkmann number (Br). These dimensionless numbers follow from writing the energy balance in a dimensionless form and the scale up notation for these groups reads:

$$\left[\frac{Gz}{Gz_0}\right]=\left[\frac{D}{D_0}\right]^{gz} \quad \text{and} \quad \left[\frac{Br}{Br_0}\right]=\left[\frac{D}{D_0}\right]^{br} \tag{15.4}$$

15.3 Summary of Equations Used

Scale-up rules are necessarily rather mathematical in nature. In this paragraph the extruder equations used are summarized.

The throughput of a single-screw extruder can be written as:

$$Q = \frac{1}{2}\pi^2 ND^2 H(1-a)\sin\theta\cos\theta \tag{15.5}$$

where θ is the flight angle and a is the throttle coefficient:

$$a = \frac{H^2 \Delta P \tan\theta}{6\mu(\pi ND)L} \tag{15.6}$$

The equation for the motor power in the pump zone can be written as:

$$E = \frac{(\pi ND)^2 WL}{H\sin\theta}(\cos^2\theta + 4\sin^2\theta + 3a\cos^2\theta) \tag{15.7}$$

where W is a channel width.

For use in scaling rules this equation can be simplified for screws with the same flight angle to:

$$E = \text{const} * \frac{\mu D^3 N^2 L}{H} \tag{15.8}$$

The pumping efficiency of the extruder is the ratio of the energy used for pumping the material and the total energy input into the extruder [1, 2, 7, 9].

$$R = \frac{QP}{E} \tag{15.9}$$

Thermal similarity yields from the energy balances:

$$\varrho C_p\left(\frac{\partial T}{\partial t} + v_x\frac{\partial T}{\partial x} + v_y\frac{\partial T}{\partial y} + v_z\frac{\partial T}{\partial z}\right) = \lambda\left(\frac{\partial^2 T}{\partial x^2} + \frac{\partial^2 T}{\partial y^2} + \frac{\partial^2 T}{\partial z^2}\right) + q \tag{15.10}$$

In this equation q is the heat produced by viscous dissipation:

$$\begin{aligned} q = {} & 2\mu\left\{\left(\frac{\partial v_x}{\partial x}\right)^2 + \left(\frac{\partial v_y}{\partial y}\right)^2 + \left(\frac{\partial v_z}{\partial z}\right)^2\right\} \\ & + \mu\left\{\left(\frac{\partial v_x}{\partial y} + \frac{\partial v_y}{\partial x}\right)^2 + \left(\frac{\partial v_x}{\partial z} + \frac{\partial v_z}{\partial x}\right)^2 + \left(\frac{\partial v_y}{\partial z} + \frac{\partial v_z}{\partial y}\right)^2\right\} \end{aligned} \tag{15.11}$$

If the equations above are made dimensionless there remain two important dimensionless numbers that govern the heat balances in the extruder, the Graez number and the Brinkmann number.

$$Gz = \frac{UH^2}{aL} \quad \text{and} \quad Br = \frac{\mu U^2}{\lambda \Delta T} \tag{15.12}$$

where λ is thermal conductivity, T the temperature, $U = \pi ND$.

The Graez number accounts for the development of the temperature profile, while the Brinkmann number signifies the ratio between viscous dissipation and heat conduction to the wall.

15.4 Kinematic Similarity

Kinematic similarity means equal shear levels in the small and the large extruder [7]. Its importance is coupled to the requirements for:

- equal mixing in small and large machines,
- equal distribution of viscous dissipation,
- equal influence of non-Newtonian rheological effects.

For the throughput of the small laboratory extruder we can write:

$$Q_0 = \frac{1}{2}\pi^2 N_0 D_0^2 H_0 (1-a_0) \sin\theta_0 \cos\theta_0 \tag{15.13}$$

and for the throughput of the production machine:

$$Q = \frac{1}{2}\pi^2 N D^2 H (1-a) \sin\theta \cos\theta \tag{15.14}$$

If the screws of the small and the large machine have the same screw angle, which is the same as the same dimensionless pitch we may write:

$$\frac{Q}{Q_0} = \frac{N D^2 H}{N_0 D_0^2 H_0} \frac{(1-a)}{(1-a_0)} \tag{15.15}$$

and if we process both machines with the same throttle coefficient:

$$\frac{Q}{Q_0} = \frac{N}{N_0}\left(\frac{D}{D_0}\right)^2 \frac{H}{H_0} \tag{15.16}$$

Introducing the diameter ratios as defined before:

$$\left(\frac{D}{D_0}\right)^q = \left(\frac{D}{D_0}\right)^n \left(\frac{D}{D_0}\right)^2 \left(\frac{D}{D_0}\right)^h = \left(\frac{D}{D_0}\right)^{n+2+h} \tag{15.17}$$

gives the exponent equation:

$$q = n + 2 + h \tag{15.18}$$

Because both machines operate with the same throttle coefficient:

$$a = a_0 \rightarrow \frac{H^2 \Delta P}{6\mu(\pi N D) L} \tan\theta = \frac{H_0^2 \Delta P_0}{6\mu_0(\pi N_0 D_0) L_0} \tan\theta_0 \tag{15.19}$$

and equal throttle coefficients leads to:

$$2h + p - \upsilon - 1 - n - \ell = 0 \tag{15.20}$$

For equal velocity gradients an extra equation is necessary:

$$\frac{\pi ND}{H} = \text{constant}$$

and therefore:

$$h = n + 1 \tag{15.21}$$

for kinematic similarity both Equations (15.19) and (15.20) must be valid:

$$p = \ell - h + \upsilon \tag{15.22}$$

These results have to be combined with geometrical considerations or with thermal scaling rules.

15.5 Geometrical and Kinetic Similarity

Geometrical similarity is often used for its simplicity but it is not a strong requirement. Especially in processing vegetable raw materials, where temperature and temperature homogeneity are very important, the principle of geometric similarity of small and large scale equipment cannot always be retained. Geometric similarity means that all dimensions scale in the same way, or:

$$l = 1 \quad \text{and} \quad h = 1 \tag{15.23}$$

Geometric and kinematic similarity follow from a combination of this equation with Equations 15.18, 15.21 and 15.22 resulting in

$$n = 0; \qquad q = 3 \quad \text{and} \quad p = \upsilon \tag{15.24}$$

This means that if we scale our extrusion cooking process in a single-screw extruder and the temperature profiles are of lesser importance:

- the rotation speed must remain the same
- the throughput should be kept proportional to D^3
- the die should be designed such that the pressure ratio equals the ratio between the end viscosities.

In this case no consideration is given to the temperature development.

15.6 Motor Power and Torque

The motor power in the extruder can be approximated to:

$$E = \text{const} * \frac{\mu D^2 N^2 L}{H} \tag{15.25}$$

It should be realised that this equation does not comprise the power needed to transport the solid bed, however, this is not important for the thermal considerations in the next paragraphs.

The scale factor of the motor power can be defined as:

$$\frac{E}{E_0} = \left(\frac{D}{D_0}\right)^e \tag{15.26}$$

and we find:

$$e = 3 + 2n + \ell + \upsilon - h \tag{15.27}$$

and for the torque:

$$m = 3 + n + \ell + \upsilon - h \tag{15.28}$$

15.7 Equal Average End Temperature

Two types of thermal similarities can be used: equal average end temperatures and similar temperature profiles [5, 6, 12]. The concept of equal average end temperatures can be applied if the extruder operates adiabatically or if $Br \gg 1$. In this case scaling up has to proceed according to equal motor power per unit throughput:

$$\frac{E}{Q} = \text{const} \quad \text{or} \quad e - q = 0 \tag{15.29}$$

With Equation 15.20 this leads, for equal viscosities and die pressures ($\upsilon = 0$ and $p = 0$), to

$$2h = 1 + n + l \tag{15.30}$$

In this case various degrees of freedom are still retained.

15.8 Similar Temperature Profiles

From the dimensionless energy equation it follows that thermal similarity can be attained if the dimensionless numbers of Graez and Brinkmann are the same for both sizes of machines.

Because:

$$Br = \frac{\mu(\pi N D)^2}{\lambda \Delta T} \tag{15.31}$$

we find for materials with the same heat conductivity (λ) that thermal similarity is attained if:

$$\upsilon + 2n + 2 = 0 \tag{15.32}$$

This means that for materials with the same viscosity: $n = -1$.

From:

$$Gz = \frac{\pi NDH^2}{aL} \tag{15.33}$$

follows at equal heat diffusivity a:

$$1 + n + 2h - \ell = 0 \tag{15.34}$$

leading to thermal similarity (equal Br and Gz numbers) if:

$$2h = \ell \tag{15.35}$$

For extruders with equal length to diameter ratios ($\ell = 1$) the channel depth must decrease according to $h = {}^1/_2$ which gives together with Equation 15.18:

$$q = 2 + n + h = 1.5 \tag{15.36}$$

or:

$$\frac{Q}{Q_0} = \left[\frac{D}{D_0}\right]^{1.5} \tag{15.37}$$

From an economical point of view this is very unfavorable, and should only be applied in very special situations.

15.9 Similarity in Residence Times

Equal residence time can be achieved if the volume divided by the throughput remains constant, or, if we define Z as the average residence time [4, 8]:

$$Z = \text{const}\frac{HLW}{Q} \tag{15.38}$$

which yields for screws with equal helix angle:

$$z = h + 1 + l - q \tag{15.39}$$

or with:

$$q = 2 + n + h \tag{15.40}$$

we find that:

$$z = l - n - 1 \tag{15.41}$$

For screws with geometric similarity, this means that ($l = -1, h = 1$ and $z = -$n), equal residence times are only possible if the rotation speed is constant. In other cases equal residence times can only be obtained by changing the screw length, according to:

$$l = 1 + n \tag{15.42}$$

Table 15.1 Application of Equation 15.30, giving a variety of possibilities for scaling rules.

n	h	q
−1.0	0.5	1.5
−0.6	0.7	2.1
−0.4	0.8	2.4
0.0	1.0	3.0

15.10
Guidelines for Scaling

In extrusion-cooking generally both heat of conduction and heat of dissipation are important in the process. In small machines the Brinkmann number is relatively small but in larger machines the dissipation becomes more dominant and the process becomes more adiabatic. Because the thermal problems are predominant the basis for the guide lines are Equation 15.30, this equation can be combined with various other (less strict) requirements. Application of Equation 15.30 gives a variety of possibilities for scaling rules. The results for screws with equal length to diameter ratio, for instance, are shown in Table 15.1).

Equal end temperatures with adiabatic operation still leave the degrees of freedom to scale according to similar temperature profiles (of course!) with $q = 1.5$ or to scale kinematically with $q = 3$ and with values in between. For the design of extrusion-cookers this means that the thermal stability of the material and of the process are important. It can be envisioned that, for the compounding process for the preparation of starch, kinematic scale-up is preferred because temperature effects are still mildly important but kinematic similarity is important to obtain the same mixing mechanism (and therefore the same material) in the small and large scale process [8, 11]. On the other hand, for processes like extrusion of snack pellets, thermal similarity is extremely important, leading to thermal scale-up. Profile extrusion and sheet extrusion are "in between" processes and could be designed with $n = -0.4$ and $h = 0.8$ leading to $q = 2.4$.

In the examples above the L/D ratio remains constant but the screw length can also be changed to retain extra degrees of freedom. This leads to a three-dimensional matrix of parameters, but is outside the scope of this book.

References

1 Brouwer, T., Todd, D.B., and Janssen, L.P.B.M. (1998) Drag and pressure flow with special twin screw mixing elements. Proceedings of the Polymer Processing Society, North American Meeting, Toronto CDN, August, 17–19, 30–31.

2 Bruin, S., van Zuilichem, D.J., and Stolp, W. (1978) A review of fundamental and engineering aspects of extrusion of biopolymers in single-screw extruder. *J. Food Proc. Eng.*, **2**, 1–37.

3 van der Goffard, D., Wal, D.J., Klomp, E.M., Hoogstraten, H.W., Janssen, L.P.B.M., Breysse, L., and Trolez, Y. (1996) Three-dimensional flow modelling of a self-wiping co-rotating twin-screw extruder, Part l. The transporting section. *Pol. Eng. Sci.*, **36**, 901–911.

4 de Graaf, R.A., Rohde, M., and Janssen, L.P.B.M. (1997) A novel model predicting the residence time distribution during reactive extrusion. *Chem. Eng. Sci.*, **52**, 4345–4356.

5 Harper, J.M. (1981) *Extrusion of Foods*, CRC Press Inc., Boca Raton, Florida.

6 Janssen, L.P.B.M., Moscicki, L., and Mitrus, M. (2002) Energy balance in food extrusion-cooking. *Int. Agrophys.*, **16** (3), 191–195.

7 Janssen, L.P.B.M., Rozendal, P.F., Hoogstraten, H.W., and Cioffi, M. (2001) A dynamic model for multiple steady states in reactive extrusion. *Int. Polym. Proc.*, **XV**, 263–271.

8 Janssen, L.P.B.M., (1998) On the stability of reactive extrusion. *Polym. Eng. Sci.*, **38**, 2010–2019.

9 Tsao, T.F., Harper, J.M., and Repholz, K.M. (1978) The effects of screw geometry on extruder operational characteristics. *AIChE Symp. Ser.*, **7** (172), 142.

10 Wal, D.J., van der Goffard, D., Klomp, E.M., Hoogstraten, H.W., and Janssen, L.P.B.M. (1996) Three-dimensional flow modelling of a self-wiping co-rotating twin-screw extruder, Part II. The kneading section. *Pol. Eng. Sci.*, **36**, 912–924.

11 Yacu, W.A. (1983) Modelling of a two-screw co-rotating extruder, in *Thermal Processing and Quality of Foods*, Elsevier Applied Science Publishers, London.

12 van Zuilichem, D.J. (1992) Extrusion Cooking. Craft or Science? PhD thesis, Wageningen University, The Netherlands.

16
Producers of Food Extruders and Expanders

Leszek Mościcki

The history of food extruders goes back to the 1950s while the production peak of this type of machinery occurred in the last 20 years. Although uncomplicated autogenic, single-screw extruders are still available on the market, their application seems to be constantly declining. Today's modern food extruders are equipped with auxiliary installations facilitating the control, maintenance and operation which, in turn, extends the options of their application. The units that have gained recognition by modern manufacturers are multipurpose extruders of modular design fitted with replaceable sets of screws and/or screw elements. The modular character of today's extruders enables operators to set up their own configurations that will meet current production needs.

The growing requirements of food producers urge machinery designers to offer better solutions not only for food extruders but also for auxiliary devices whose performance gives the opportunity of manufacturing a broad assortment of extrusion-cooked food and feed.

Tables 16.1 and 16.2 contain basic data on the majority of the world extruder and expander manufacturers including a short description. The data provided have been collected from the literature on the subject and the information available in the public domain as well as the author's personal experience. The data covers the period 2004–2009. Regrettably, not all information is currently updated by equipment producers, especially the smaller ones. Most of them have in recent years probably modified their types of series or stopped production of some machinery (usually the least complex). The markings or designations of some machines have also changed. By presenting the list below, I intended to familiarize the reader with the scope and profile of production of the leading manufacturers of extruders and expanders by simply enumerating them. This will certainly help you contact the manufacturers and find the machinery that will best suit your needs.

As with many manufacturers of food machinery, the extruder market also has its pioneers that lead the world in this new field of the agri-food industry. They not only offer extruders but also comprehensive processing lines. Most companies, however, focus on the manufacturing of specific types of extruders or expanders, offering the

Extrusion-Cooking Techniques: Applications, Theory and Sustainability. Edited by Leszek Moscicki

ISBN: 978-3-527-32888-8

Table 16.1 Technical specification of the equipment.

Producer	Model	Single screw	Twin-screw Co-rotating	Twin-screw Counter-rotating	Modular screws	Diameter (mm)	Heating			Capacity ($kg\,h^{-1}$)	Power. (kW)
							Oil	Water/ steam	Electric		
1	2	3	4	5	6	7	8	9	10	11	12
ALMEX	AL 400	*			*	400		*		7000–15 000	400–500
	AL 350	*			*	350		*		4000–9000	315–400
	AL 300	*			*	300		*		2500–8000	132–315
	AL 200	*			*	200		*		1500–3000	75–132
	AL 150	*			*	150		*		500–1200	45–75
AMANDUS KAHL	OE38	*			*	360				18 000–28 000	250–315
	OE30	*			*	275				10 000–18 000	160–250
	OE23	*			*	230				500–800	110–160
	OE15	*			*	151				1000–2500	45–75
ANDERSON	EXPANDER 4-5″	*				114				10–200	18
INTERNATIONAL	EXTRUDER 6″	*				152				200–500	37
	COOKER 8″	*			*	203				500–4000	37–112
	10″	*				254				4000–9000	112–373
	12″	*				305				3600–9100	110–260
ANDRITZ SPROUT	617	*			*	177		*	*	1000–5000	250
	620	*			*	210		*	*	2000-7000	250
	917	*			*	177		*	*	2000–6000	315
	920	*			*	210		*	*	2000–10 000	315
	1250	*			*	250		*	*	8000–20 000	600
BAKER PERKINS	MPF 30		*			30		*	*	37–135	7,5
	MPF 40		*			40		*	*	75–275	15
	MPF 50	*	*		*	50	*	*	*	160–560	28

	MPF 65	*	*		*	65	*	*	*	300–1100	60
	MPF 80		*		*	80	*	*	*	600–2200	115
	MPF 100		*		*	100	*	*	*	1200–4400	225
	MPF 125		*		*	125	*	*	*	2400–7600	350
	MPF 160		*		*	160	*	*	*	3200–12 000	600
	MP 2019		*			19		*		5–20	2
	BPF 100							*		200	6,7
	BPF 200							*		1000	30
BERGA	ES/PF 25	*			*	225		*		1000–3000	120
	ESC 3	*			*	226		*		1000–3000	120
BRABENDER OHG	DSE 35/12 D			*		35			*	30–40	8,5
	LAB M.19/20 DN	*				19			*	5	3,3
BÜHLER AG	BASF	*			*	133	*	*		3000–4000	110
	BASH	*			*	178	*	*		3000–6000	150
	BCTL-42/BCTC-10		*		*	42	*	*		50–450	110
	BCTG-62/BCTC-22		*		*	62	*	*		150–1400	216
	BCTF-93/BCTC-48		*		*	93	*	*		500–4500	754
	BCTH-125/BCTC-100		*		*	125	*	*		1000–10 000	900
	BCTJ-175/BCTC-160		*		*	175	*	*		2000–20 000	900
	ECOtwin™ 125		*		*	125	*	*		10 000	450
	ECOtwin™ 175		*		*	175	*	*		20 000	900
BUSS	LR 300	*			*	300	*	*		600–15 000	280
	LR 250	*			*	260	*	*		300–9000	240
	LR 200	*			*	200	*	*		200–4000	115
	LR 100	*			*	100	*	*		30–800	50
CLEXTRAL	BC 21		*	*	*	25	*	*	*	5–100	2,2–9
	BC 45		*		*	55,5	*	*	*	50–700	15–16
	BC 72		*		*	88	*	*	*	200–3000	48–226
	BC 82		*		*	102	*	*	*	300–4500	71–310

(*Continued*)

Table 16.1 (Continued)

Producer	Model	Single screw	Twin-screw Co-rotating	Twin-screw Counter-rotating	Modular screws	Diameter (mm)	Heating			Capacity (kg h^{-1})	Power. (kW)
							Oil	Water/ steam	Electric		
	BC 92		*		*	115	*	*	*	500–6000	98–453
	BC 105		*		*	132	*	*	*	1000–9000	138–645
	BC 160		*		*	200	*	*	*	4000–16 000	416–2060
EXTRUTECH	E-325	*			*	83		*		100	22
	E-525	*			*	133		*		250–3000	75
	E-750	*			*	191		*		2000–16 000	149
	E-925	*			*	235		*		3000–28 000	187
FEN-FOOD	G (EXTRUDERS)	*			*	100	*			100–3000	20–150
ENTERPRISE	F (FORMERS)	*			*	130	*				
		*			*	160	*				
		*			*	190	*				
		*			*	210	*				
FUDEX GROUP	2FB60		*		*	60			*	70–200	36
	2FB 90		*		*	85			*	200–600	100
	2FB105		*		*	105			*	300–1000	200
	2FB135		*		*	135			*	800–2500	300
INOTEC	INOTEX 50	*			*	88		o	*	350–450	37
INTERNATIONAL	INOTEX 100	*			*	131		o	*	800–950	75
	INOTEX 125	*			*	131		o	*	1100–1400	90
	INOTEX 125 S	*			*	131			*	1400	75–90
	INOTEX 200	*			*	175		o	*	2200–3000	167–240
	INOTEX 200 S	*			*	175			*	3000–3800	167–240
	INOQUICK 2300 E	*			*	252			*	10 000–15 000	200–240

INSTAPRO	9800	*		*			*	2700–4500	263
	9600	*		*			*	2700–6300	263
	2500	*		*			*	1100–1500	90
	2000 DB	*		*			*	1176–3490	56–112 (x2)
	(Double Barrel)								
	2000 RC	*		*			*	457–726	56
	600	*		*			*	272–365	PTO/37
LALESSE	DS. 80		*	*	80		*	100–400	65
	90 E	*			75		*	100	22
	200 E	*			110		*	200	35
	LA 85	*			85		*	100	30
LE MECCANICA	LAMEC 305	*			305			10 000–15 000	160–250
NAMSUNG	NS 130	*		*	130			300–1000	75
INDUSTRIAL	FX 100		*		100		o	500–2000	111
CO. LTD.	FX 60	*	*		62		o	150–600	45
	FX 40	*	*		44		o	50–150	15
	NSE 100				134		o	300–1400	75
	NSE 50				90		o	300–400	38
PAVAN GROUP	G 55	*			55		*	10–30	5
	G 70	*		*	70		*	30–60	15
	G 90	*	*	*	90	*	*	80–120	18
	G 130	*	*	*	130	*	*	250–350	45
	G 150	*	*	*	160	*	*	450–550	75
	G 180	*	*	*	180	*	*	600–800	90
	G 200	*	*	*	200	*	*	800–1000	160
	F 55	*	*		55	*		20–40	4
	F 70	*		*	70	*	*	45–80	5,5
	F 90	*		*	90	*	*	90–140	15
	F 130	*		*	130		*	300–400	30

(Continued)

Table 16.1 (Continued)

Producer	Model	Single screw	Twin-screw Co-rotating	Twin-screw Counter-rotating	Modular screws	Diameter (mm)	Heating			Capacity ($kg\,h^{-1}$)	Power. (kW)
							Oil	Water/ steam	Electric		
	F 150	*			*	150			*	500–600	45
	F 175	*			*	175			*	900–1000	75
	F 200	*			*	200			*	1300–1400	110
	F 240	*				240				1800–2200	170
	TT 40					40				100–200	34
	TT 58					58				200–400	100
	TT 70					70				500–800	180
	TT 92					92				600–2000	325
	TT 112					112				1200–4000	500
	TT 133					133				2400–8000	730
PLANET	SYSTEM 100	*			*	60		*		100	13–25
FLOWLINE											
TELEDYNE	CP 2	*	*			51	*	*	*	1–100	3–900
READCO	CP 5	*	*			127	*	*	*	100–1000	3–900
	CP 8	*	*			203	*	*	*	200–340	3–900
	CP 10	*	*				*	*	*	340–78	3–900
	CP 12	*	*			381	*	*	*	780–1600	3–900
	CP 15	*	*			457	*	*	*	1600–5400	3–900
	CP 18	*	*				*	*	*	5400–9500	3–900
	CP 20	*	*			610	*	*	*	9500–12 800	3–900
	CP 24	*	*			760	*	*	*	12 800–22 000	3–900
	CP 30	*	*				*	*	*	22 000–43 200	3–900
WENGER	X-85/4	*			*	85	*	*		200–800	30

	X-115/7	*			*	115	*	*		500–2000	95
	X-165/32	*	*		*	165	*	*		1500–6000	150
	X-185/54	*	*		*	216	*	*		3500–14 000	350
	X-235/108	*	*		*	235	*	*		5000–20 000	450
	X-285/108	*	*		*	285	*	*		11 000–25 000	500
	TX 144 MAGNUM		*		*	144	*	*		800–7500	300
	TX 115 MAGNUM		*		*	115	*	*		350–6800	
	TX 85 MAGNUM				*	85	*	*		150–2800	112
	TX 57 MAGNUM				*	57	*	*		50–850	
	C^2TX-8.1				*		*	*		1000–8000	300
	C^2TX-16.2				*		*	*		2000–16 000	600
WERNER and	ZSK 177		*		*	177	*	*	*	1000–12 000	2500
PFLEIDERER	ZSK 133		*		*	133	*	*	*	800–7000	1330
–COPERION	ZSK 92		*	*	*	92	*	*	*	400–3000	440
	ZSK 70		*		*	70	*	*	*	200–1000	240
	ZSK 58		*	*	*	58	*	*	*	50–500	130
	ZSK 40		*		*	40	*	*	*	20–120	45
	ZSK 25		*	*	*	25	*	*	*	5–40	9,5
	CONTINUA 37		*		*	37	*	*		10–50	10
	CONTINUA 58		*		*	58	*	*	*	50–200	37
	CONTINUA 83		*		*	83	*	*	*	200–1000	185
	CONTINUA 120		*		*	120	*	*	*	500–4000	455
	CONTINUA 170		*		*	170	*	*	*	1000–8000	960
	CONTINUA 240		*		*	240	*	*		1500–14 000	1700
ZMCH	TS 45	*				45			*	25	14
METALCHEM	TS 60	*				60			*	50	25
GLIWICE	2S 9/5			*		90/50	*	*	*	200	55

* standard equipment, o optional equipment.

Table 16.2 Additional information.

Producer	Model	Modular barrel	Barrel with axial opening	Variable rpm	Constant rpm	Feed	Food	Additional information concerning the offer
1	2	3	4	5	6	7	8	9
ALMEX	AL 500	*		*	*	*		
	AL 350	*		*	*	*		Complete lines for pet food and aquafeed
	AL 300	*		*	*	*		
	AL 200	*		*	*	*		
	AL 150	*		*	*	*		
AMANDUS KAHL	OE 38	*			*	*		Expanders and complete lines for pet food and aquafeed
	OE 30	*			*	*		
	OE 23	*			*	*		
	OE 15	*			*	*		
ANDERSON INTERNATIONAL	EXPANDER 4-5″ EEC			*	*	*	*	Complete lines for pet food and aquafeed
	EXTRUDER 6″ EEC			*	*	*	*	
	COOKER 8″ EEC	*		*	*	*	*	
	COOKER 10″ EEC	*		*	*	*	*	
	COOKER 12″ EEC	*			*			
ANDRITZ SPROUT	617	*		*	*	*	*	Complete lines for pet food and aquafeed
	620	*		*	*	*	*	
	917	*		*	*	*	*	
	920	*		*	*	*	*	
	1250	*		*	*	*	*	
BAKER PERKINS	MPF 30	*	*	*		*	*	Complete lines for food and feed products
	MPF 40	*	*	*		*	*	
	MPF 50	*	*	*		*	*	
	MPF 65	*	*	*		*	*	

	MPF 80	*	*	*		*	*	
	MPF 100	*	*	*		*	*	
	MPF 125	*	*	*		*	*	
	MPF 160	*	*	*		*	*	
	MPF 2019	*	*	*		*		
	BPF 100			*				
	BPF 200			*				
BERGA	ES/PF 25	*			*	*	*	Dryers and expanders
	ESC 3	*			*	*	*	
BRABENDER OHG	DSE 35/12 D			*		*	*	Laboratory equipment
	LAB M.19/20 DN			*				
BÜHLER AG	BASF	*		*	o	*	*	Complete lines for food and feed products (full range)
	BASH	*		*	o	*	*	
	BCTL-42/BCTC-10	*		*	o	*	*	
	BCTG-62/BCTC-22	*		*	o	*	*	
	BCTF-93/BCTC-48	*		*	o	*	*	
	BCTH-125/BCTC-100	*		*	o	*	*	
	BCTJ-175/BCTC-160	*		*	o	*	*	
	ECOtwin™ 125	*		*	o	*	*	
	ECOtwin™ 175	*		*	o	*	*	
BUSS	LR 300	*	*	*		*	*	Extruders
	LR 250	*	*	*		*	*	
	LR 200	*	*	*		*	*	
	LR 100	*	*	*		*	*	
CLEXTRAL	BC 21	*		*		*	*	Complete lines for food and feed products (full range)
	BC 45	*		*		*	*	
	BC 72	*		*		*	*	

(Continued)

Table 16.2 (Continued)

Producer	Model	Modular barrel	Barrel with axial opening	Variable rpm	Constant rpm	Feed	Food	Additional information concerning the offer
	BC 82	*		*		*	*	
	BC 92	*		*		*	*	
	BC 105	*		*		*	*	
	BC 160	*		*		*	*	
EXTRUTECH	E-325	*		*		*	*	Complete lines for food and feed products
	E-525	*		*		*	*	
	E-750	*			*	*	*	
	E-925	*			*	*	*	
FEN FOOD	G (EKSTRUDERY)	*	*	*			*	Complete lines for snack food
ENTERPRISE	F (FORMERY)							
FUDEX GROUP	2 FB 60	*	*	*			*	Baby food and crispbread lines
	2 FB 90	*	*	*			*	
	2 FB 105	*	*	*			*	
	2 FB 135	*	*	*			*	
INOTEC	INOTEX 50	*		o	*	*	*	Dies, extruders, dryers
INTERNATIONAL	INOTEX 100	*		o	*	*		
	INOTEX 125	*		o	*	*		
	INOTEX 125 S	*			*	*		
	INOTEX 200	*		o	*	*		
	INOTEX 200 S	*		o	*	*		
	INOQUICK 2300 E	*			*			
INSTAPRO	9800	*		o	*	*	*	Complete lines for feed components
	9600	*		o	*	*	*	
	2500	*		o	*	*	*	
	2000 DB (Double Barrel)	*		o	*	*	*	
	2000 RC	*		o	*	*	*	

	600	*		o	*	*	*	
LALESSE	DS. 80	*		*			*	Complete lines for direct extruded snacks
	90 E			*			*	
	200 E			*			*	
LE MECCANICA	LAMEC 305	*			*	*		Feed equipment
NAMSUNG	NS 130	*	*		*	*		Feed and food equipment
INDUSTRIAL,	FX 100	*		o	*	*	*	
CO. LTD.	FX 60	*	*	o	*	*	*	
	FX 40	*		*	*	*	*	
	NSE 100	*		o	*	*	*	
	NSE 50	*		o	*	*	*	
PAVAN GROUP	G 55	*		*			*	Complete lines for food products (full range), packaging equipment
	G 70	*		*			*	
	G 90	*		*			*	
	G 130	*		*			*	
	G 150	*		*			*	
	G 180	*		*			*	
	G 200	*		*			*	
	F 55			*			*	
	F 70	*		*			*	
	F 90	*		*			*	
	F 130	*		*			*	
	F 150			*			*	
	F 175			*			*	
	F 200			*			*	
	F 240			*			*	
	TT 40			*			*	
	TT 58			*			*	

(Continued)

Table 16.2 (*Continued*)

Producer	Model	Modular barrel	Barrel with axial opening	Variable rpm	Constant rpm	Feed	Food	Additional information concerning the offer
	TT 70			*			*	
	TT 92			*			*	
	TT 112			*			*	
	TT 133			*			*	
PLANET FLOWLINE	SYSTEM 100	*	*	*			*	Complete lines for direct extruded snacks
TELEDYNE	CP 2	*	*	*	*	*	*	Feed equipment
READCO	CP 5	*	*	*	*	*	*	
	CP 8	*	*	*	*	*	*	
	CP 10	*	*	*	*	*	*	
	CP 12	*	*	*	*	*	*	
	CP 15	*	*	*	*	*	*	
	CP 18	*	*	*	*	*	*	
	CP 20	*	*	*	*	*	*	
	CP 24	*	*	*	*	*	*	
	CP 30	*	*	*	*	*	*	
WENGER	X-85/4	*		*	*	*		Complete lines for food and feed products (full range)
	X-115/7	*		*	*	*		
	X-165/32	*	*	*	*	*	*	
	X-185/54	*	*	*	*	*	*	
	X-235/108	*	*	*	*	*	*	
	X-285/108	*	*	*	*	*	*	
	TX 144 MAGNUM	*	*	*		*	*	
	TX 115 MAGNUM	*	*	*		*	*	
	TX 85 MAGNUM	*		*		*		
	TX 57 MAGNUM	*		*		*		

	C^2TX-8.1	*		*		*		
	C^2TX-16.2	*		*		*		
WERNER and PFLEIDERER – COPERION	ZSK 177	*		*		*	*	Specializes in production of wide range of equipment for food
	ZSK 133	*		*		*	*	
	ZSK 92	*	*	*		*	*	
	ZSK 70	*		*		*	*	
	ZSK 58	*		*		*	*	
	ZSK 40	*		*		*	*	
	ZSK 25	*		*		*	*	
	CONTINUA 37	*		*		*	*	
	CONTINUA 58	*		*		*	*	
	CONTINUA 83	*		*		*	*	
	CONTINUA 120	*		*		*	*	
	CONTINUA 170	*		*		*	*	
	CONTINUA 240	*		*		*	*	
ZMCH	TS 45			*	o	o	*	Extruders
METALCHEM	TS 60			*	o	o	*	
GLIWICE	2S 9/5			*	o	*	*	

* standard equipment, o optional equipment.

auxiliary equipment as a complementary option for complete processing lines. A few of the largest producers, such as Buhler AG, Clextral, Pavan Group. or Wenger Inc. offer a full range of devices for the production of various extrudates and specialize in one field, for example, breakfast cereals, or both fields – food and feed – simultaneously.

When selecting the supplier of machinery, it is necessary first to define future assortment to be manufactured and to perform a detailed market analysis for the prospective application of a purchased device or line. A universal device frequently proves to be a bad choice and economizing on machinery may have an adverse effect. The automation of the production process may definitely improve line efficiency but involves extra expenditure (minimum 10% of the total machinery cost). Because the machines in question are expensive, a thorough pre-purchase analysis is definitely recommended and justified, even though the manufacture of extrusion-cooked food is regarded as highly profitable.

Index

Extrusion-Cooking Techniques: Applications, Theory and Sustainability. Edited by Leszek Moscicki

ISBN: 978-3-527-32888-8

v

w